Measure & Probability

S R Athreya

Indian Statistical Institute
Bangalore

V S Sunder

Institute of Mathematical Sciences
Chennai

CRC Press
Taylor & Francis Group
Boca Raton London New York

CRC Press is an imprint of the
Taylor & Francis Group, an **informa** business

Dedicated, with respect,
to the memory of
Paul Halmos

Contents

Preface   vii

1   Probabilities and Measures   1

    1.1   Introduction   1
    1.2 σ-Algebras as Events   2
    1.3   Algebras, Monotone Classes, etc.   6
    1.4   Preliminaries on Measures   8
    1.5   Outer Measures and Caratheodory Extension   12
    1.6   Lebesgue Measure   18
    1.7   Regularity   21
    1.8   Bernoulli Trials   23

2   Integration   26

    2.1   Measurable Functions   26
    2.2   Integration   29
    2.3   a.e. Considerations   38

3   Random Variables   42

    3.1   Distribution and Expectation   42
    3.2   Independent Events and Tail σ-Algebra   45
    3.3   Some Distributions   47
    3.4   Conditional Expectation   52

4   Probability Measures on Product Spaces   56

    4.1   Product Measures   56
    4.2   Joint Distribution and Independence   61

4.3   Probability Measures on Infinite Product Spaces   62
4.4   Kolmogorov Consistency Theorem   67

5   Characteristics and Convergences   73

5.1   Characteristic Functions   73
5.2   Modes of Convergence   79
5.3   Central Limit Theorem   88
5.4   Law of Large Numbers   92

6   Markov Chains   101

6.1   Discrete Time MC   101
6.2   Examples   104
6.3   Classification of States   108
6.4   Strong Markov Property   123
6.5   Stationary Distribution   128
6.6   Limit Theorems   131

7   Some Analysis   137

7.1   Complex Measures   137
7.2 L^p Spaces   142
7.3   Radon–Nikodym Theorem   148
7.4   Change of Variables   156
7.5   Differentiation   159
7.6   The Riesz Representation Theorem   168

A   Appendix   182

A.1   Metric Spaces   182
A.2   Topological Spaces   184
A.3   Compactness   192
A.4   The Stone–Weierstrass Theorem   205

B   Tables   215

References   217
Index   219

Preface

An initial set of notes, essentially containing three of the first four chapters of this book, was prepared when such a course was given by the second author at the 'Mathematics Training and Talent Search Programme' at IIT, Mumbai in 1993. These notes were re-used subsequently in courses given to students at the 'NURTURE' programme at the Institute of Mathematical Sciences (IMSc), as well as in a 'core course' on real analysis at IMSc. Later, when the second author had to give a variant of this course to beginning graduate students at the Chennai Mathematical Institute, some 'analysis', meaning some basic material about L^p spaces, absolute continuity, etc., seemed to be called for; this was the genesis of parts of the last chapter of this book.

When it was subsequently suggested that there should be a serious 'probability component' to the book for it to be useful, the first author was approached with the plea that he might help modify the book so that this lacuna might be rectified to some extent. Chapters 3, 5 and 6 were the result. The 'Appendix' was included almost as a second thought, with the hope that it may be a source for the 'uninitiated student' to fill in some possible gaps in the prerequisites needed for reading this book. This appendix has been 'lifted' verbatim from an appendix (written for similar reasons) of the book [28]. The authors plead guilty to having yielded to the temptation of the ease and convenience of the 'cut-and-paste option' offered by the word-processing software; more importantly, they are very grateful to Rajendra Bhatia and the TRIM series for kindly granting the permission necessary to so 'lift' this material.

It is hoped that this book could be read by anybody with a bachelor's degree in mathematics. Even this 'requirement' is not necessary if the prospective reader is blessed with a 'modicum of mathematical maturity'; an appendix has been supplied for precisely such a reader. We believe the entire contents of this book could be very comfortably covered in a two-semester course, while a reasonable one-semester course can be fashioned by selecting appropriate parts of the book according to the need of the student (or taste of the instructor).

The first chapter sets as its goal the construction of Lebesgue measure. After a brief discussion on the need for restricting oneself to a suitable class of 'measurable sets', we get down to definitions and basic properties of abstract σ-algebras (as well as algebras and monotone classes) of subsets of a given universal set Ω, and pass on to measures defined on such algebras. The fundamental Caratheodory extension theorem is then stated and proved. Finally the existence (and uniqueness) of Lebesgue measure is deduced, and the existence of 'non-Lebesgue-measurable' sets demonstrated.

The second chapter begins by discussing measurable functions, and then establishes the fundamental proposition regarding approximability of positive measurable functions by simple functions. We then go on to define the 'Lebesgue' integral of appropriate functions, and derive such basic results as the monotone and dominated convergence theorems, Fatou's lemma, etc. The chapter ends with a brief discussion of the notion 'almost everywhere'.

The third chapter introduces the reader to the probabilistic terminology and approach: to start with, the terminology of a random variable (on an abstract probability space) and its distribution (the 'push-forward' probability measure defined on the Borel subsets of the real line) are discussed. The crucial notion of '(stochastic) independence' of events (or random variables) is introduced, and the Borel-Cantelli Lemma and Kolmogorov Zero-One Law' proved. Some of the more standard distributions , both discrete (Bernoulli, Binomial, Poisson, etc.) and continuous (Uniform, Normal, etc.), are described. And a final section is devoted to conditional expectations and probabilities.

The common link running through the fourth chapter is 'measures on product spaces'. Section 4.1 proves the existence of products of σ-finite measures on arbitrary spaces, and goes on to prove the fundamental theorems of Tonelli and Fubini. The next section connects product measures and independent random variables through the notion of their 'joint-distribution' (an appropriate 'push-forward measure'). Section 4.3 is devoted to (vacuously) applying the Caratheodory extension theorem to construct interesting examples of probability measured on arbitrary products of finite sets, and paves the way for Markov chains (at least in the case of finite state space). Section 4.4 is a quick discussion of Kolmogorov's consistency theorem, and the exercises here introduces the reader to 'transition probability measures' and 'Markov processes' in general as well as the 'standard Brownian motion' in particular.

The fifth chapter is devoted to proving two cornerstones of probability theory, viz., the Central Limit Theorem and the Law of Large Numbers. The first two sections pave the way with preliminary results such as uniqueness and inversion theorem for 'characteristic functions (or Fourier transforms) of distributions', and theorems of Slutsky, Skorohod, Polya and Scheffe on relations between various modes of convergence. Section 5.3 proves the Central Limit Theorem (for i.i.d. random variables with finite variance), after preparing the ground with the continuity theorem which lists reformulations of convergence in distribution. The final Section, 5.4, derives the Strong Law of Large Numbers from an auxiliary result to the effect that the 'sample means' of a sequence of stationary random variables with finite moment converge almost surely to the conditional expectation with respect to the 'invariant σ-algebra'.

The sixth chapter focuses on discrete-time Markov chains on countable state spaces. We begin with an introduction to the basic notions of aperiodicity, irreducibility, transience, recurrence and stationarity. We then proceed to prove the two main limit theorems in the area, namely convergence to stationarity for aperiodic irreducible chains and the renewal theorem. We also explicitly exhibit, following [11], the stationary measure for an irreducible finite state space Markov chain. We conclude with a discussion of recurrence and transience properties for birth-death and queuing chains.

The seventh chapter addresses various topics typically covered in early graduate courses in analysis. Section 7.1 addresses 'finite complex measures' (as well as finite real measures, traditionally called signed measures), and establishes that such measures are necessarily of 'finite total variation' and are expressible as linear combinations of finite positive measures. Section 7.2, devoted to L^p-spaces, establishes Hölder's inequality and the fact that L^p-spaces are Banach spaces. The Radon–Nikodym theorem is proved (following von Neumann) in Section 7.3., and then used to prove the duality among L^p-spaces corresponding to conjugate indices as well as the Hahn decomposition of signed measures and Lebesgue–Nikodym theorem. Section 7.4 is a brief digression into the 'change of variables formula', while Section 7.5 establishes classical results such as the fundamental theorem of calculus (for absolutely continuous functions). Finally, Section 7.6 is concerned with the Riesz Representation Theorem which is stated and proved in three flavours: the compact metric case, the general compact Hausdorff case, and finally the locally compact Hausdorff case.

Most of the topics covered here are 'standard fare', but several proofs of 'standard theorems' are possibly unusual and not too 'standard'. This comment might apply to our treatment of the following results: the Strong Law of Large Numbers (see Theorem 5.4.4), the Hahn decomposition of signed measures (Proposition 7.3.7), and the Riesz Representation Theorem (Theorems 7.6.1, 7.6.7, 7.6.9). To be entirely honest, however, we learnt later that V.S. Varadarajan had also constructed a proof of the Riesz Representation Theorem which, like ours, is based on the Hahn–Banach Theorem. Unfortunately, that proof appeared long ago and in a not very easily located journal. Similarly, our proof of SLLN was based on a lecture by Michael Keane which the first author attended, and the basic idea of the proof can be found in [18].

It is a pleasure to record here our gratitude to the Institute of Mathematical Sciences and Indian Statistical Institute for the wonderful facilities and atmosphere they have been providing us for the last several years. In particular, we would also like to thank Rajeeva Karandikar, S. Kesavan, Rahul Roy, K. Ramamurthy and S. Ramasubramanian for various conversations, and S. Sundar for pointing out loopholes in the earlier versions of some of the proofs.

Siva R. Athreya
V.S. Sunder
January 11, 2008

1

Probabilities and Measures

1.1 Introduction

Measure theory is all about computing lengths, areas, volumes, etc. All these attributes—length (of a linear set), area (of a planar set) and volume (of a set in 3-dimensional space)—are what are known as measures defined for (suitable) subsets of an ambient space—usually denoted by Ω (which is 1-, 2- or 3-dimensional space in the above examples).

An advantage of this abstraction is that the abstract notion can be applied to diverse situations, ranging from 0-dimensional space (the set of possible outcomes of a random experiment with only finitely many possible outcomes, where the 'measure' of a set of possible events is the 'probability of that event') to finite-dimensional spaces $\mathbb{R}^n$, $n \geq 1$, where the celebrated 'Lebesgue measure' is the n-dimensional volume, and even infinite-dimensional spaces (such as $C([0, \infty))$, where the Wiener measure of appropriate sets of functions is the probability that a random path executing Brownian motion belongs to that set).

Just to set the ball rolling, we first consider the natural choice for the 'sample space' Ω when one wishes to consider probabilistic questions regarding some simple experiments with different possible outcomes.

Experiment 1: Toss a fair coin twice and note down the outcome of each toss. We may describe the collection of all (four) possible outcomes by $\Omega = \{HH, HT, TH, TT\}$, where H is head and T is tail.

Experiment 2: Roll a six-faced fair die and note the outcome. We know that there are six possible outcomes: $\Omega = \{1, 2, 3, 4, 5, 6\}$.

Experiment 3: Toss a fair coin till the first head appears and note down the number of tosses made. We may parametrise the outcomes of this experiment by the set $\Omega = \mathbb{N}$ of natural numbers.

Experiment 4: Pick a random point 'uniformly distributed in $I = (0, 1]$' and note down the outcome. Here the outcome space is $\Omega = I$.

More general experiments will lead to various outcome sets Ω. Given an experiment with the set of all outcomes denoted by Ω, one is interested in specific events. For example, in Experiment 1, let E_1 be the event {the first toss results in H}; in Experiment 2, let E_2 be the event {the outcome is an even number}; in Experiment 3, let E_3 be the event {head appears in at most 10 tosses} and in Experiment 4, let E_4 be the event {the number chosen is bigger than one-half}.

As can be seen above, each such event is typically a subset of Ω. In Experiments 1, 2 and 3, we were able to enumerate the set of outcomes, so we could take the set of all subsets of Ω, the power set $\mathcal{P}(\Omega)$, as the set of all events; however, when the set Ω is infinite, as in Experiment 4, there is a problem with being able to define the probability of arbitrary subsets of Ω. In order to understand precisely what the nature of this problem is, consider the following question.

Question: Given a set $S \subset \mathbb{R}^3$, can we measure the volume of S?

Almost by definition, the volume of the unit cube in $\mathbb{R}^3$ is 1 cubic unit. From this starting point, we would like to systematically calculate volumes of more complicated solids.

Let us investigate 'reasonable requirements' to be met by the process of assigning volumes to sets. Specifically, if $|S|$ denotes the volume of the set $S \subseteq \mathbb{R}^3$, it appears natural to demand that the assignment $S \mapsto |S|$ must satisfy:

(1) $|S| \in [0, \infty]$. (Volumes cannot be negative.)
(2) $|S_1 \cup S_2| = |S_1| + |S_2|$ if $S_1 \cap S_2 = \emptyset$. (Volumes of non-overlapping sets should 'add up'.)
(3)   If $\{S_n\}_{n=1}^{\infty}$ is an increasing sequence of sets with union S, i.e., $S_1 \subset S_2 \subset \ldots$ and $S = \bigcup_{n=1}^{\infty} S_n = S$, then $|S| = \lim_{n \to \infty} |S_n|$. (This is a 'continuity requirement' which seems natural.)
(4) $|C| = 1$, if C is a cube of side 1.
(5)   If $x \in \mathbb{R}^3$ and $S + x = \{s + x : s \in S\}$ denotes the translate of S by x, then $|S + x| = |S|$: (volumes remain invariant under translation.)

Unfortunately, it is impossible to define $|S|$ for all subsets S of $\mathbb{R}^3$ in such a way that conditions (1)–(5) above are all satisfied. This problem is caused by the fact that there are some very strange 'non-measurable' sets.

Hence, whether the motivation comes from computing probabilities of certain events of an experiment having outcome space Ω or computing lengths of subsets of a space Ω, the concept of events or subsets needs to be sorted out first. We do precisely that in the next section.

1.2 σ-Algebras as Events

Let Ω denote a set. Depending on the context, you can think of it as the set of all possible outcomes of some experiment or as any arbitrary set where we want to understand concepts such as length, area or volume.

For any subset E of Ω, let us write E' for its complement.

Definition 1.2.1   *A collection $\mathcal{B}$ of subsets of a set Ω is called a σ-algebra if*

(1) $\Omega \in \mathcal{B}$;

(2) $E \in \mathcal{B} \Rightarrow E' \in \mathcal{B}$ *and*

(3)   *if $E_1, E_2, \ldots, E_n, \ldots$ is a sequence in $\mathcal{B}$, then $\cup_{n=1}^{\infty} E_n \in \mathcal{B}$.*

Condition (3) above is paraphrased as: 'σ-algebras are closed under countable unions'.

Remark 1.2.2   *If $\mathcal{B}$ is a σ-algebra of subsets of Ω, then observe that:*

(1) $\emptyset(= \Omega') \in \mathcal{B}$ *and*

(2)   *if $E_n \in \mathcal{B}, n = 1, 2, \ldots.$, then $\cap_{n=1}^{\infty} E_n \left(= (\cup_{n=1}^{\infty} E_n')'\right) \in \mathcal{B}$.*

Thus a σ-algebra is closed under the formation of complements, countable unions and countable intersections.

The following proposition provides various examples of σ-algebras. It also introduces the key idea of when a σ-algebra is to be regarded as being generated by a collection of sets S.

Proposition 1.2.3

(1)   *If $\{\mathcal{B}_i : i \in I\}$ is any collection of σ-algebras of subsets of Ω, then $\cap_{i \in I} \mathcal{B}_i = \{E \subseteq \Omega : E \in \mathcal{B}_i \; \forall i \in I\}$ is also a σ-algebra.*

(2) $2^\Omega = \{E \subseteq \Omega\}$ *is a σ-algebra.*

(3)   *If S is any collection of subsets of Ω, then there exists a smallest σ-algebra of subsets of Ω which contains S; in fact, this minimal σ-algebra is denoted by $\sigma(S)$ and is given by $\cap\{\mathcal{B} : \mathcal{B}$ is a σ-algebra of subsets of Ω such that $S \subseteq \mathcal{B}\}$; we shall call it the σ-algebra generated by S.*

Proof: See Ex. 1.2.5.

As suggested in Proposition 1.2.3, we may take $\mathcal{B} = \mathcal{P}(\Omega) \equiv 2^\Omega$ in Experiments 1, 2 and 3 above. We will see that this sets up a 'probability structure' on $\{\Omega, \mathcal{B}\}$; however, we will see before the end of this chapter that this is not the case for Experiment 4, where we shall, instead, take $\mathcal{B} = \sigma((a, b] : a, b \in (0, 1])$; and that all the requirements (1)–(5) listed in Section 1.1 will be satisfied *provided* we take care to demand that we shall make sense of $|S|$ only when $S \in \mathcal{B}$. The term *measurable* is often suggestively used to indicate the sets for which we 'can speak sensibly of their volume'.

We shall eventually prove the following result:

Theorem 1.2.4 *Let S denote the collection of all n-dimensional parallelopipeds. Thus, a typical element of S is of the form $B = \prod_{i=1}^{n}[a_i, b_i] = \{x = (x_1, \ldots, x_n) \in \mathbb{R}^n : a_i \le x_i \le b_i \text{ for } 1 \le i \le n\}$, where a_i, b_i are arbitrary real numbers satisfying $a_i \le b_i$.*
   Let $B = \sigma(S)$ denote the σ-algebra generated by S.
   *Then there exists a unique function $\mu : B \to [0, \infty]$ (called **Lebesgue measure** on $\mathbb{R}^n$) satisfying the following:*

(1) $\mu\left(\prod_{i=1}^{n}[a_i, b_i]\right) = \prod_{i=1}^{n}(b_i - a_i)$;

(2) *if $\{E_n\}_{n=1}^{\infty} \subseteq B$ and $E_n \cap E_m = \emptyset$ for $n \ne m$, then $\mu\left(\bigcup_{n=1}^{\infty} E_n\right) = \sum_{n=1}^{\infty} \mu(E_n)$.*

   Further, if $E \in B$ and $x \in \mathbb{R}^n$, then $E + x \in B$ and $\mu(E + x) = \mu(E)$.

Note that if $n = 3$ and C is a rectangular cuboid, its volume $| C |$ agrees with $\mu(C)$ defined in (1) above.

EXERCISES

Ex. 1.2.5   Prove Proposition 1.2.3.

Ex. 1.2.6   The following experiment is performed: a die is rolled and then a coin is tossed. Outcomes of the experiment are noted. Describe the outcome space Ω and the event space (or suitable σ-algebra) B for this experiment.

Ex. 1.2.7   The curriculum vitae of two male applicants for a college teaching position in probability are placed in the same file as the curriculum vitae of two female applicants. Two positions become available and the first, at the rank of professor, is filled by selecting one of the four applicants at random. The second position, at the rank of assistant professor, is then filled by selecting at random one of the three remaining applicants. List

(1) the elements of the outcomes space Ω;

(2) the elements of the event A that the position of the assistant professor is filled by a male applicant;

(3) the elements of the event B that exactly one of the two positions is filled by a male applicant;

(4) the elements of the event C that neither position is filled by a male applicant.

(5) Also, sketch a Venn diagram to show the relationship among the events A, B, C and Ω.

Ex. 1.2.8 Consider the sample space $\Omega = \{a, b, c, d, e\}$. Given that $\{a, b, e\}$, and $\{b, c\}$ are both events (or, in other words, elements of the σ-algebra, what other events are implied by taking unions, intersections and complements?

Ex. 1.2.9 Assume $\mathcal{B}$ is a σ-algebra on Ω. Let $B \subset \Omega$. Let $C = \mathcal{B} \cup \{B\}$. Show that $\sigma(C) = \{(B \cap U) \cup (B' \cap V) : U, V \in \mathcal{B}\}$.

Ex. 1.2.10 Assume $\mathcal{B}$ is a σ-algebra on Ω. If $\{A_1, A_2, \ldots A_n\}$ is a partition of Ω, then describe $\sigma(\mathcal{B} \cup \{A_1, A_2, \ldots, A_n\})$.

Ex. 1.2.11 Let $\mathcal{B}_1, \mathcal{B}_2$ be two σ-algebras on Ω. Show that $\sigma(\mathcal{B}_1 \cup \mathcal{B}_2) = \sigma(C)$, where $C = \{A_1 \cup A_2 : A_1 \in \mathcal{B}_1, A_2 \in \mathcal{B}_2\}$.

Ex. 1.2.12 Let $n \in \mathbb{N}$. The σ-algebra $\mathcal{B}^n$ on $\mathbb{R}^n$ generated by open subsets of $\mathbb{R}^n$ is called the **Borel** σ**-algebra of** $\mathbb{R}^n$, and the members of $\mathcal{B}^n$ are called Borel sets.

(1) For $B \subset \mathbb{R}$ and $x \in \mathbb{R}$, let $x + B = \{x + y : y \in B\}$. Show that for every element $B \in \mathcal{B}$, $x + B$ is also a Borel set (i.e., an element of the Borel σ-algebra).

(2) Let $T : \mathbb{R} \to \mathbb{R}$ be defined as $T(x) = x + 1$. Show that $\widetilde{\mathcal{B}} = \{B \in \mathcal{B} : TB = B\}$ is a σ-algebra on $\mathbb{R}$.

(3) If (X, d) is a metric space and if $\mathcal{S}_1 = \{B(x, r) : x \in X, r > 0\}$ is the set of open balls in X and $\mathcal{S}_2$ is the set of open sets in X, verify that $\sigma(\mathcal{S}_1) = \sigma(\mathcal{S}_2)$ provided X is a separable metric space. In that case, this common σ-algebra is called the **Borel** σ**-algebra** of X. It is denoted by $\mathcal{B}_X$, and its members are called Borel sets. Also, if $f : X \to X$ is a homeomorphism, then show that $E \in \mathcal{B}_X \Rightarrow f(E) \in \mathcal{B}_X$.

Ex. 1.2.13 Let $\mathcal{B}$ be the Borel σ-algebra on $\mathbb{R}$. Define $\mathcal{S}_1 = \{(a, b) : -\infty < a < b < \infty\}$, $\mathcal{S}_2 = \{[a, b] : -\infty < a < b < \infty\}$, $\mathcal{S}_3 = \{[a, b) : -\infty < a < b < \infty\}$, $\mathcal{U}$ is the collection of all open sets in $\mathbb{R}$, $\mathcal{C}$ is the collection of all closed sets in $\mathbb{R}$ and $\mathcal{K}$ is the collection of all compact sets in $\mathbb{R}$. Show that $\mathcal{B} = \sigma(\mathcal{S})$, where $\mathcal{S}$ is any one of the collections $\mathcal{S}_1, \mathcal{S}_2, \mathcal{S}_3, \mathcal{U}, \mathcal{C}$ or $\mathcal{K}$.

Ex. 1.2.14 Let $\Omega = (0, 1]$ and $\mathcal{B}$ be the σ-algebra generated by open sets in Ω, i.e., the Borel σ-algebra on Ω. Show that $\widetilde{\mathcal{B}} = \{B \subset \Omega : B \in \mathcal{B}$ and is either disjoint from $(\frac{1}{2}, 1]$ or contains $(\frac{1}{2}, 1]\}$ is a σ-algebra on Ω. Moreover, $\widetilde{\mathcal{B}} = \sigma$(collection of all intervals contained in $(0, \frac{1}{2})$).

Ex. 1.2.15 Let $\mathcal{B}^n$ be the Borel σ-algebra in $\mathbb{R}^n$ for $n \in \mathbb{N}$.

(1) Show that $\mathcal{B}^1 = \sigma((a, b] : -\infty \leq a \leq b \leq \infty)$.

(2)   Show that $\mathcal{B}^2 = \sigma((a, b] \times (c, d] : -\infty \le a \le b \le \infty, -\infty \le c \le d \le \infty)$.

(3)   Show that $\mathcal{B}_1 = \{B \times \mathbb{R} : B \in \mathcal{B}^1\}$ and $\mathcal{B}_2 = \{\mathbb{R} \times B : B \in \mathcal{B}^1\}$ are both σ-algebras on $\mathbb{R}^2$ contained in $\mathcal{B}^2$ and that $\mathcal{B}^2 = \sigma(\mathcal{B}_1 \cup \mathcal{B}_2)$.

(4)   Let $T : \mathbb{R}^2 \to \mathbb{R}^2$ be the map $(x, y) \to (y, x)$. Show that $\tilde{\mathcal{B}} = \{B \in \mathcal{B}^2 : TB = B\}$ is a σ-algebra and find a generating set.

1.3   Algebras, Monotone Classes, etc.

Even for understanding just σ-algebras of sets, we will need the auxiliary notions of algebras and monotone classes of sets. We continue to let Ω denote a general set (e.g., $\mathbb{R}^n$, or any of the outcome spaces discussed earlier). Symbols denoted by script letters such as $\mathcal{A}, \mathcal{B}, \mathcal{M}, \mathcal{S}$, etc., will always represent classes of subsets of Ω.

Definition 1.3.1

(1)   *A family $\mathcal{A}$ is said to be an* **algebra** *if* :

    (a) $\Omega \in \mathcal{A}$;
    (b) $E \in \mathcal{A} \Rightarrow E' \in \mathcal{A}$ *and*
    (c) $E, F \in \mathcal{A} \Rightarrow E \cup F \in \mathcal{A}$.

(2)   *A family $\mathcal{M}$ is said to be a* **monotone class** *provided:*

    (a)   *If $E_1 \subseteq E_2 \subseteq E_3 \subseteq, \ldots$ and $E_n \in \mathcal{M}$ $\forall n$, then $\cup_{n=1}^{\infty} E_n \in \mathcal{M}$; and*
    (b)   *If $E_1 \supseteq E_2 \supseteq E_3, \ldots$ and if $E_n \in \mathcal{M}$ $\forall n$, then $\cap_{n=1}^{\infty} E_n \in \mathcal{M}$.*

Thus $\mathcal{A}$ is an algebra if it is closed under complements and finite unions, and, hence, under finite intersections (since $E \cap F = (E' \cup F')'$); while $\mathcal{M}$ is a monotone class if it is closed under forming 'monotone limits'. If the E_n are as in Definition 1.3.1 (2a) (resp., (2b)), we shall write $E_n \uparrow \cup_{m=1}^{\infty} E_m$ (resp., $E_n \downarrow \cap_{m=1}^{\infty} E_m$).

Proposition 1.3.2 is popularly referred to as the Monotone Class Lemma. It gives a description of the σ-algebra generated by an algebra $\mathcal{A}$ of sets. (It turns out to be precisely the monotone class $\mathcal{M}(\mathcal{A})$ generated by $\mathcal{A}$; this last notion is defined entirely analogously to $\sigma(\mathcal{A})$—see Ex. 1.3.4 below for the definition and notation used.)

Proposition 1.3.2   (the Monotone class lemma) *If $\mathcal{A}$ is an algebra of subsets of Ω, then $\mathcal{M}(\mathcal{A}) = \sigma(\mathcal{A})$.*

Proof: It is enough to check that $\mathcal{M}(\mathcal{A})$ is closed under complements and unions. (Can you see why?)

As for complements, check that if

$$\mathcal{M}_c = \{E \in \mathcal{M}(\mathcal{A}) : E' \in \mathcal{M}(\mathcal{A})\},$$

then $\mathcal{A} \subseteq \mathcal{M}_c$ and $\mathcal{M}_c$ is a monotone class; hence, $\mathcal{M}_c = \mathcal{M}(\mathcal{A})$ and it follows that $\mathcal{M}(\mathcal{A})$ is closed under formation of complements.

Concerning unions, verify first that if $A \in \mathcal{A}$, and if

$$\mathcal{M}_A = \{E \in \mathcal{M}(\mathcal{A}) : E \cup A \in \mathcal{M}(\mathcal{A})\},$$

then $\mathcal{M}_A \supseteq \mathcal{A}$ ($\mathcal{A}$ is closed under unions); since $\mathcal{M}_A$ is clearly a monotone class, conclude that $\mathcal{M}_A = \mathcal{M}(\mathcal{A})$. Then, if $A \in \mathcal{M}(\mathcal{A})$ and $\mathcal{M}_A$ is defined as above, note that we have shown that $\mathcal{M} \supset A$, where

$$\mathcal{M} = \{A \in \mathcal{M}(\mathcal{A}) : \mathcal{M}_A = \mathcal{M}(\mathcal{A})\}$$
$$= \{A \in \mathcal{M}(\mathcal{A}) : E \in \mathcal{M}(\mathcal{A}) \Rightarrow E \cup A \in \mathcal{M}(\mathcal{A})\}.$$

The latter description of $\mathcal{M}$ shows that $\mathcal{M}$ is a monotone class; in other words, $\mathcal{M}(A)$ is closed under unions, and the proof is complete.   □

EXERCISES

Ex. 1.3.3 If $\Omega \in \mathcal{A}$ and if $A, B \in \mathcal{A}$ implies that $A \cap B' \in \mathcal{A}$, then show that $\mathcal{A}$ is an algebra.

Ex. 1.3.4

(1) If $\mathcal{B}$ is a σ-algebra, then $\mathcal{B}$ is an algebra as well as a monotone class.
(2) If $\mathcal{B}$ is an algebra as well as a monotone class, then $\mathcal{B}$ is a σ-algebra.
(3) Imitate Proposition 1.2.3 and show that if $\mathcal{S}$ is any class of subsets of Ω, then there exists a smallest algebra (resp., monotone class) containing $\mathcal{S}$. This algebra (resp., monotone class), which will henceforth be denoted by $\mathcal{A}(\mathcal{S})$ (resp., $\mathcal{M}(\mathcal{S})$) is given by $\mathcal{A}(\mathcal{S}) = \cap\{\mathcal{A} : \mathcal{A} \supseteq \mathcal{S}, \ \mathcal{A}$ is an algebra of subsets of $\Omega\}$ (resp., $\mathcal{M}(\mathcal{S}) = \cap\{\mathcal{M} : \mathcal{M} \supseteq \mathcal{S}, \ \mathcal{M}$ is a monotone class of subsets of $\Omega\}$.

Ex. 1.3.5

(1) How many distinct algebras of subsets of Ω exist if Ω is a three-element set?
(2) If $\mathcal{A}$ is a finite algebra of subsets of (a possibly infinite set) Ω, what can you say about the number of distinct sets in $\mathcal{A}$?

Ex. 1.3.6 If $\mathcal{A}$ is an algebra of sets in Ω, and if $\{E_n\}_{n=1}^{\infty} \subseteq \mathcal{A}$, then there exist sequences $\{S_n\}_{n=1}^{\infty}, \{D_n\}_{n=1}^{\infty}$ of sets in $\mathcal{A}$ such that:

(1) $S_1 \subset S_2 \subset S_3 \subset \ldots$
(2) $D_n \cap D_m = \emptyset$ if $n \neq m$ and
(3) $\cup_{k=1}^{n} E_k = \cup_{k=1}^{n} S_k = \cup_{k=1}^{n} D_k$, for $n = 1, 2, \ldots$

Further, conditions (1)–(3) uniquely determine the sets S_n and D_n for all n.

Ex. 1.3.7 Let $\Omega = \mathbb{R}$, let S denote the collection of intervals of the form $(a,b]$, where $-\infty \leq a < b \leq \infty$ (where of course $(a, \infty]$ is to be interpreted as (a, ∞)). Show that a typical non-empty element of $A(S)$, the algebra generated by S is of the form $\coprod_{k=1}^{n} I_k$, where $n = 1, 2, \ldots$, and $I_k \in S$ for $1 \leq k \leq n$.

Ex. 1.3.8 Let A be an algebra of subsets of a set Ω. Let $\Omega_0 \subseteq \Omega$. Define $A \cap \Omega_0 = \{A \cap \Omega_0 : A \in A\}$. Show that $A \cap \Omega_0$ is an algebra of subsets of Ω_0, and that $A \cap \Omega_0$ is a σ-algebra if A is one. (In the case when $\Omega_0 \in A$, it is more natural and customary to write $A|_{\Omega_0}$ instead of $A \cap \Omega_0$.) What can you say if A is a monotone class?

1.4   Preliminaries on Measures

We will now begin to construct the final piece in the puzzle: i.e., build the function μ of Theorem 1.2.4 or the probability of one of our experiments. Throughout this section, A will denote an algebra of subsets of a fixed (universal) set Ω.

Definition 1.4.1

(1)   A **set function** defined on A is a function $\mu : A \to [0, \infty]$;
(2)   a set function $\mu : A \to [0, \infty]$ is said to be **finitely additive** if $A, B \in A$, $A \cap B = \emptyset \Rightarrow \mu(A \cup B) = \mu(A) + \mu(B)$.

Before proceeding further, let us agree to use the symbol $\coprod$ to denote 'disjoint union'. Thus, we shall write $A \coprod B$ to denote $A \cup B$ if A and B are disjoint sets; similarly, we shall write $\coprod A_i$ to denote $\cup A_i$ if $\{A_i\}$ is a collection of pairwise disjoint sets—i.e., $A_i \cap A_j = \emptyset$ if $i \neq j$.

Thus a finitely additive set function is one which satisfies $\mu(A \coprod B) = \mu(A) + \mu(B)$.

The reason for the terminology 'finitely additive' stems from the following proposition.

Proposition 1.4.2 μ *is a finitely additive set function if and only if*

$$\mu\left(\coprod_{i=1}^{n} A_i\right) = \sum_{i=1}^{n} \mu(A_i)$$

for A_i in A; i.e., if $A_1, \ldots, A_n \in A$ and $A_i \cap A_j = \emptyset$ for $i \neq j$, then $\cup_{i=1}^{n} A_i \in A$ and $\mu\left(\cup_{i=1}^{n} A_i\right) = \sum_{i=1}^{n} \mu(A_i)$.

Proof: Use mathematical induction.   □

Proposition 1.4.3 lists some elementary facts about finitely additive set functions.

Proposition 1.4.3   *Suppose μ is a finitely additive set function defined on $\mathcal{A}$. Then, for all $A, B \in \mathcal{A}$, we have:*

(1) $\mu(A \cup B) = \mu(A) + \mu(B \setminus A)$, *where $B \setminus A = B \cap A' = \{b \in B : b \notin A\}$; in particular, μ is sub-additive, i.e., $\mu(A \cup B) \leq \mu(A) + \mu(B)$.*

(2) $A \subseteq B \Rightarrow \mu(A) \leq \mu(B)$.

(3)   *More generally than in (1), if $A_1, \ldots, A_n \in \mathcal{A}$, then (an easy induction argument shows that)*

$$\mu\left(\cup_{i=1}^n A_i\right) \leq \sum_{i=1}^n \mu(A_i).$$

Proof: See Ex. 1.4.9.   □

It follows from (2) of Proposition 1.4.3 that $\mu(\emptyset) \leq \mu(A)$ for all $A \in \mathcal{A}$; on the other hand, finite additivity implies that $\mu(\emptyset) = \mu(\emptyset \sqcup \emptyset) = \mu(\emptyset) + \mu(\emptyset)$, which shows that $\mu(\emptyset) = 0$ or ∞; the latter case is uninteresting because then $\mu(A) = \infty$ for all $A \in \mathcal{A}$.

At the other end of the spectrum, note that $\mu(A) \leq \mu(\Omega)$ for all $A \in \mathcal{A}$ and so $\mu(A) < \infty$ for all $A \in \mathcal{A}$ if $\mu(\Omega) < \infty$.

Definition 1.4.4

(1)   *A set function μ defined on an algebra $\mathcal{A}$ is called a **measure** if:*

(a) $\mu(\emptyset) = 0$; *and*

(b) μ *is countably additive: i.e., if $\{A_n\}_{n=1}^\infty$ is a countable collection of sets in $\mathcal{A}$ such that (i) $A_n \cap A_m = \emptyset$ for $n \neq m$, and (ii) $A = \bigsqcup_{n=1}^\infty A_n \in \mathcal{A}$, then $\mu(A) = \sum_{n=1}^\infty \mu(A_n)$.*

(2)   *The measure μ is said to be a **finite measure** if $\mu(\Omega) < \infty$.*

(3)   *The measure μ is said to be a **probability measure** if $\mu(\Omega) = 1$.*

Remark 1.4.5   *Note the crucial requirement (ii) in (1b) above; thus, we require that $\mu\left(\bigsqcup_{n=1}^\infty A_n\right) = \sum_{n=1}^\infty \mu(A_n)$ only if, in addition to the A_n's being a pairwise disjoint sequence of sets in $\mathcal{A}$, it is the case that the infinite union $\cup_{n=1}^\infty A_n$ also belongs to $\mathcal{A}$. Of course, this condition is automatically met if $\mathcal{A}$ is actually a σ-algebra (and, hence, closed under countable unions). Usually, it is assumed that the domain of definition of a measure is a σ-algebra and not just an algebra. We shall adopt the more general definition given above until we have established the fundamental extension theorem of*

Caratheodory which states that any measure defined on an algebra $\mathcal{A}$ admits a unique extension to a measure defined on the generated σ-algebra $\sigma(\mathcal{A})$. In fact, the main reason for introducing Sections 1.2–1.4 is to lead up to a proof of that theorem.

Proposition 1.4.6 *Suppose μ is a measure defined on an algebra $\mathcal{A}$ of subsets of Ω. Then,*

(1) μ is **continuous from below***: i.e., if $A_n \in \mathcal{A}$, $A_1 \subseteq A_2 \subseteq \ldots$ and $A = \bigcup_{n=1}^{\infty} A_n \in \mathcal{A}$, then $\mu(A) = \lim_{n\to\infty} \mu(A_n)$;*

(2) *if $\mu(\Omega) < \infty$, then μ is **continuous from above***: i.e., if $A_n \in \mathcal{A}$, $A_1 \supseteq A_2 \supseteq \ldots$ and $A = \bigcap_{n=1}^{\infty} A_n \in \mathcal{A}$, then $\mu(A) = \lim_{n\to\infty} \mu(A_n)$.*

Proof: Let $\{D_n\}_{n=1}^{\infty}$ be 'the disjointification' of $\{A_n\}_{n=1}^{\infty}$ as in Ex. 1.3.6.
Thus,

(1) $D_n \in \mathcal{A}$ and $A_n = \coprod_{k=1}^{n} D_k$, for $n = 1, 2, \ldots$. Then clearly, $A = \coprod_{k=1}^{\infty} D_k$, and by the assumed countable additivity of μ, we find that

$$\mu(A) = \sum_{k=1}^{\infty} \mu(D_k)$$

$$= \lim_{n\to\infty} \sum_{k=1}^{n} \mu(D_k)$$

$$= \lim_{n\to\infty} \mu(A_n)$$

(2)   Note that for any $E \in \mathcal{A}$, $\mu(\Omega) = \mu(E) + \mu(E')$, and since $\mu(E') \leq \mu(\Omega) < \infty$, we find:

$$\mu(E) = \mu(\Omega) - \mu(E').$$

Now, if $A_n \downarrow A$, clearly then $A_n' \uparrow A'$: so that, by (1),

$$\mu(A) = \mu(\Omega) - \mu(A')$$

$$= \mu(\Omega) - \lim_{n\to\infty} \mu(A_n')$$

$$= \lim_{n\to\infty} (\mu(\Omega) - \mu(A_n'))$$

$$= \lim_{n\to\infty} \mu(A_n) \qquad \square$$

Finiteness of the measure is a necessary condition for continuity from above (see Ex. 1.4.10). We conclude this section by observing that a measure is necessarily **countably sub-additive**.

Proposition 1.4.7 *If μ is a measure defined on an algebra $\mathcal{A}$ of subsets of Ω, if $A_1, A_2, \ldots \in \mathcal{A}$, and $\cup_{n=1}^{\infty} A_n \in \mathcal{A}$, then $\mu \left(\cup_{n=1}^{\infty} A_n \right) \leq \sum_{n=1}^{\infty} \mu(A_n)$.*

Proof: Let $\{S_n\}_{n=1}^{\infty}$ be the increasing version of $\{A_n\}_{n=1}^{\infty}$ as in Ex. 1.3.6; thus $S_n = \cup_{k=1}^{n} A_k$, for $n = 1, 2, \ldots$.

Then $S_n \in \mathcal{A}$ and $\cup_{n=1}^{\infty} S_n = A$; hence, by Proposition 1.4.6(1) and Proposition 1.4.3(3),

$$\mu(A) = \lim_{n \to \infty} \mu(S_n)$$

$$\leq \lim_{n \to \infty} \sum_{k=1}^{n} \mu(A_k)$$

$$= \sum_{n=1}^{\infty} \mu(A_n). \qquad \qquad \Box$$

EXERCISES

Ex. 1.4.8 Suppose μ is a finitely additive set function defined on $\mathcal{A}$.

(1) Show that $\mu(A \cup B) + \mu(A \cap B) = \mu(A) + \mu(B) \; \forall A, B \in \mathcal{A}$. In particular, if $\mu(\Omega) < \infty$ (so that there is no problem with subtraction), then

$$\mu(A \cup B) = \mu(A) + \mu(B) - \mu(A \cap B).$$

(2) Can you write down a general formula for $\mu(A \cup B \cup C)$, $A, B, C \in \mathcal{A}$, (or, more generally, for $\mu \left(\cup_{i=1}^{n} A_i \right)$ when $\{A_i\}_{i=1}^{n} \subseteq \mathcal{A}$), under the assumption that $\mu(\Omega) < \infty$?

Ex. 1.4.9 Prove Proposition 1.4.3.

Ex. 1.4.10 Let Ω be a countable set and $\mathcal{A} = 2^{\Omega}$ be the collection of all subsets of Ω. Then,

(1) $\mathcal{A}$ is a σ-algebra.
(2) If $\mu : \mathcal{A} \to [0, \infty]$ is defined by $\mu(E) =$ number of elements in E, then μ is a measure and is called **the counting measure on** Ω (since μ counts the number of elements in a set).
(3) If $\Omega = \{w_1, w_2, \ldots\}$ is an enumeration of Ω, and if Ω is infinite, let $A_n = \{w_n, w_{n+1}, \ldots\}$; notice that $A_n \downarrow \emptyset$ but

$$\mu(\emptyset) = 0 \neq \infty = \lim_{n \to \infty} \mu(A_n).$$

(4)   If μ is a general, possibly infinite, measure defined on an algebra $\mathcal{A}$ of
subsets of any set Ω, and if $A, A_n \in \mathcal{A}, A_1 \supseteq A_2 \supseteq \dots, A = \cap_{n=1}^{\infty} A_n$,
then (imitate the proof of Proposition 1.4.6(2) to) show that $\mu(A) = \lim_{n \to \infty} \mu(A_n)$ provided there exists some k such that $\mu(A_k) < \infty$.

1.5   Outer Measures and Caratheodory Extension

We assume throughout this section that μ is a measure defined on an algebra $\mathcal{A}$
of subsets of a fixed set Ω.

In attempting to extend μ to a larger class of sets than $\mathcal{A}$, we imitate the
process of taking 'upper Riemann sums' in the theory of Riemann integration.

Definition 1.5.1   *If μ is a measure defined on an algebra $\mathcal{A}$ of subsets of Ω,
define*

$$\mu^*(E) = \inf \left\{ \sum_{n=1}^{\infty} \mu(A_n) : E \subseteq \cup_{n=1}^{\infty} A_n, A_n \in \mathcal{A} \; \forall n \right\}$$

for all subsets E of Ω. The set function μ^* is called the **outer measure** *associated
with the measure μ.*

Our strategy for proving the extension theorem will be to show that:

(1) $\mu^*|_{\mathcal{A}} = \mu$ and that
(2) $\mu^*|\sigma(\mathcal{A})$ is a measure.

We make the following modest first step in that direction.

Proposition 1.5.2

(1) $\mu^*|\mathcal{A} = \mu$; *i.e.,* $\mu^*(A) = \mu(A)$ *for all* $A \in \mathcal{A}$ *and*
(2) μ^* *is countably sub-additive; i.e.,* $\mu^* \left(\cup_{n=1}^{\infty} E_n \right) \leq \sum_{n=1}^{\infty} \mu^*(E_n)$ *for all*
$E_1, E_2, \dots \subseteq \Omega$.

Proof:

(1)   If $A \in \mathcal{A}$, then $A \subseteq A \cup \emptyset \cup \emptyset \cup \dots$ and $A, \emptyset \in \mathcal{A}$; hence,

$$\mu^*(A) \leq \mu(A) + 0 + 0 + \dots = \mu(A).$$

Conversely, if $A, A_1, A_2, \dots \in \mathcal{A}$ and $A \subseteq \cup_{n=1}^{\infty} A_n$, then also $\{A \cap A_n\}_{n=1}^{\infty} \subseteq \mathcal{A}$ and $A = \cup_{n=1}^{\infty} (A \cap A_n)$; the countable sub-additivity and
monotonicity of μ imply that

$$\mu(A) \le \sum_{n=1}^{\infty} \mu(A \cap A_n)$$

$$\le \sum_{n=1}^{\infty} \mu(A_n);$$

the arbitrariness of the collection $\{A_n\}_{n=1}^{\infty}$ and the definition of $\mu^*(A)$ now show that $\mu(A) \le \mu^*(A)$.

(2)   Suppose $E = \cup_{n=1}^{\infty} E_n$, $E_n \subset \Omega$. Let $\varepsilon > 0$. For each n, there exists $\{A_{n,k}\}_{k=1}^{\infty} \subseteq \mathcal{A}$ such that $E_n \subseteq \cup_{k=1}^{\infty} A_{n,k}$ and $\mu^*(E_n) + \varepsilon/2^n \ge \sum_{k=1}^{\infty} \mu(A_{n,k})$; sum over n to get

$$\sum_{n=1}^{\infty} \mu^*(E_n) + \varepsilon \ge \sum_{k,n=1}^{\infty} \mu(A_{n,k})$$

$$\ge \mu^*(E)$$

since $E = \cup_{n=1}^{\infty} E_n \subseteq \cup_{n,k=1}^{\infty} A_{n,k}$. Since this is true for all $\varepsilon > 0$, we have

$$\mu^*(E) \le \sum_{n=1}^{\infty} \mu^*(E_n). \qquad \square$$

We now single out those subsets of Ω which are well-behaved with respect to the measure μ (or, equivalently, with respect to the associated outer measure μ^*).

Definition 1.5.3   *A subset $A \subseteq \Omega$ is said to be μ^*-**measurable** if*

$$\mu^*(E) = \mu^*(E \cap A) + \mu^*(E \cap A')$$

for $E \subseteq \Omega$. The collection of all μ^*-measurable sets will be denoted by $\mathcal{M}(\mu)$.

Remark 1.5.4   *Since μ^* is (countably) sub-additive (cf. Proposition 1.5.2 (2)), it is clear that a set A is μ^*-measurable if and only if*

$$\mu^*(E) \ge \mu^*(E \cap A) + \mu^*(E \cap A')$$

for all subsets E of Ω.

Proposition 1.5.5   *With the preceding notation, we have: $\mathcal{A} \subseteq \mathcal{M}(\mu)$.*

Proof: If $A \in \mathcal{A}$ and $E \subseteq \Omega$, let $\varepsilon > 0$ be arbitrarily fixed and choose $\{A_n\}_{n=1}^{\infty} \subseteq \mathcal{A}$ such that $E \subseteq \cup_{n=1}^{\infty} A_n$ and

$$\mu^*(E) + \varepsilon \ge \sum_{n=1}^{\infty} \mu(A_n).$$

Since μ is additive on $\mathcal{A}$, and since $E \cap B \subseteq \bigcup_{n=1}^{\infty}(B \cap A_n)$ for $B \varepsilon \mathcal{A}$ (in particular, for $B \varepsilon \{A, A'\}$), deduce that

$$\mu^*(E \cap A) + \mu^*(E \cap A') \leq \sum_{n=1}^{\infty} \mu(A_n \cap A) + \sum_{n=1}^{\infty} \mu(A_n \cap A')$$

$$= \sum_{n=1}^{\infty} \mu(A_n)$$

$$\leq \mu^*(E) + \varepsilon. \qquad \square$$

Lemma 1.5.6 *If $A, B \in \mathcal{M}(\mu)$ and $A \subseteq B$, then $B - A \in \mathcal{M}(\mu)$.*

The proof of Lemma 1.5.6 is elementary and is left as an exercise to the reader.

Lemma 1.5.7 *If $D = \bigsqcup_{j=1}^{n} D_j$ and $D_j \in \mathcal{M}(\mu)$ for $1 \leq j \leq n$, then $D \in \mathcal{M}(\mu)$ and*

$$\mu^*(E) = \sum_{j=1}^{n} \mu^*(E \cap D_j) + \mu^*(E \cap D')$$

for all subsets E of Ω.

Proof: We shall prove the assertion, which is obvious for $n = 1$, by induction on n.

Suppose the result is valid for $n - 1$. Then, $C = \bigsqcup_{j=2}^{n} D_j \in \mathcal{M}(\mu)$, $C' = D_1 \sqcup D'$ and

$$\mu^*(E) = \sum_{j=2}^{n} \mu^*(E \cap D_j) + \mu^*(E \cap C')$$

$$= \sum_{j=2}^{n} \mu^*(E \cap D_j) + \mu^*(E \cap C' \cap D_1) + \mu^*(E \cap C' \cap D_1')$$

$$= \sum_{j=1}^{n} \mu^*(E \cap D_j) + \mu^*(E \cap D'),$$

thus proving the lemma. $\qquad \square$

All the pieces are in place for the main result of this section.

Theorem 1.5.8 *$\mathcal{M}(\mu)$ is a σ-algebra and $\mu^*|\mathcal{M}(\mu)$ defines a measure on $\mathcal{M}(\mu)$. In particular, every measure on an algebra $\mathcal{A}$ admits a countably additive extension to the generated σ-algebra.*

Proof: It follows from Lemmas 1.5.6 and 1.5.7 that $\mathcal{M}(\mu)$ is an algebra, since $A \cup B = A \coprod (B - (A \cap B))$.

Suppose $\{A_n\}_{n=1}^{\infty} \subseteq \mathcal{M}(\mu)$; let $\{D_n\}_{n=1}^{\infty}$ be the disjointification of $\{A_n\}_{n=1}^{\infty}$: cf. Ex. 1.3.6. Thus, $\cup_{k=1}^{n} A_k = \coprod_{k=1}^{n} D_k$, and $\cup_{k=1}^{\infty} A_k = \coprod_{k=1}^{\infty} D_k$. Further, since $\mathcal{M}(\mu)$ is an algebra,

$$D_n = \left(\cup_{k=1}^{n} A_k \right) - \left(\cup_{k=1}^{n-1} A_k \right) \in \mathcal{M}(\mu).$$

Let $S_n = \coprod_{k=1}^{n} D_k = \cup_{k=1}^{n} A_k$, and $A = \cup_{n=1}^{\infty} S_n \left(= \coprod_{k=1}^{\infty} D_k = \cup_{k=1}^{\infty} A_k \right)$. Appeal to Lemma 1.5.7 (with S_n in place of D in that lemma) to find that for all $E \subseteq \Omega$,

$$\mu^*(E) = \sum_{k=1}^{n} \mu^*(E \cap D_k) + \mu^*(E \cap S_n')$$

$$\geq \sum_{k=1}^{n} \mu^*(E \cap D_k) + \mu^*(E \cap A')$$

(since $S_n \subseteq A$ implies $S_n' \supseteq A'$). Since n is arbitrary, we get

$$\mu^*(E) \geq \sum_{k=1}^{\infty} \mu^*(E \cap D_k) + \mu^*(E \cap A')$$

$$\geq \mu^*(E \cap A) + \mu^*(E \cap A')$$

$$\geq \mu^*(E)$$

(since $A = \cup_{k=1}^{\infty} D_k$, and μ^* is countably sub-additive).
Deduce that $A \in \mathcal{M}(\mu)$ and that

$$\mu^*(E) = \sum_{k=1}^{\infty} \mu^*(E \cap D_k) + \mu^*(E \cap A') \qquad (1.5.1)$$

for all $E \subseteq \Omega$.

Substitute A for E in (1.5.1) to find

$$\mu^*(A) = \sum_{k=1}^{\infty} \mu^*(D_k). \qquad (1.5.2)$$

Thus we have shown that $\mathcal{M}(\mu)$ is a σ-algebra and that $\mu^*(\coprod_{k=1}^{\infty} D_k) = \sum_{k=1}^{\infty} \mu^*(D_k)$ for arbitrary $\{D_k\}_{k=1}^{\infty} \subseteq \mathcal{M}(\mu)$. Thus the first assertion of the theorem is proved.

As for the second, $\mathcal{M}(\mu)$ is a σ-algebra containing $\mathcal{A}$, cf. Proposition 1.5.5, and so $\sigma(\mathcal{A}) \subseteq \mathcal{M}(\mu)$; the desired extension is obtained by restricting μ^* to $\sigma(\mathcal{A})$. $\qquad \square$

Remark 1.5.9   *Clearly, the requirement that μ is countably additive on $\mathcal{A}$, as defined in Definition 1.4.4(1b), is necessary for μ to admit an extension to a countably additive function on $\sigma(\mathcal{A})$. The reader is advised to find out precisely where the hypothesis of countable additivity of μ on $\mathcal{A}$ was used in proving the extension theorem. On doing so, s/he would realise that a finitely additive set function on an algebra is countably additive if and only if it is countably sub-additive.*

A natural question still remains: is the extension provided by Theorem 1.5.8 unique? In general, this is not so—see Ex. 1.5.13; however, cases of non-uniqueness are pathological. A particularly useful class of measures which do not permit such 'pathology'—as shown by Theorem 1.5.11—is identified in the definition below.

Definition 1.5.10   *A measure μ defined on an algebra $\mathcal{A}$ of subsets of a set Ω is said to be σ-**finite** if there exists a sequence $\{A_n\}_{n=1}^{\infty} \subseteq \mathcal{A}$ such that $\Omega = \cup_{n=1}^{\infty} A_n$ and $\mu(A_n) < \infty$ for all n.*

It follows from Ex. 1.3.6 that μ is σ-finite if and only if there exist sequences $\{D_n\}_{n=1}^{\infty}, \{S_n\}_{n=1}^{\infty} \subseteq \mathcal{A}$ such that $\Omega = \coprod_{n=1}^{\infty} D_n = \bigcup_{n=1}^{\infty} S_n, S_1 \subseteq S_2 \subseteq \ldots,$ and $\mu(D_n) < \infty, \mu(S_n) < \infty$ for all n.

We are now in a position to state (and prove) the Caratheodory extension theorem in its final form.

Theorem 1.5.11   **(The Caratheodory extension theorem)** *If μ is a σ-finite measure defined on an algebra $\mathcal{A}$ of subsets of a set Ω, then there exists a unique measure $\widetilde{\mu}$ defined on the generated σ-algebra $\sigma(\mathcal{A})$ such that $\widetilde{\mu}|\mathcal{A} = \mu$.*

Proof: The existence of $\widetilde{\mu}$ was already proved in Theorem 1.5.8. For uniqueness, first consider the case when μ is actually finite. If $\widetilde{\mu}_1$ and $\widetilde{\mu}_2$ are two measures defined on $\sigma(\mathcal{A})$ which extend μ, it follows from Proposition 1.4.6 that $\mathcal{M} = \{A \in \sigma(\mathcal{A}) : \widetilde{\mu}_1(A) = \widetilde{\mu}_2(A)\}$ is a monotone class. (This is where finiteness of μ is used.) Since $\mathcal{M} \supseteq \mathcal{A}$, it follows that $\mathcal{M} \supseteq \mathcal{M}(\mathcal{A}) = \sigma(\mathcal{A})$.

Now if μ is σ-finite, write $\Omega = \coprod E_n$ where $E_n \in \mathcal{A}$ and $\mu(E_n) < \infty$. If $\widetilde{\mu}_i$ are extensions of μ, observe in the notation of Ex. 1.3.8 that for each n, the restriction of μ to $\mathcal{A}|E_n$ is a finite measure, that $\sigma(\mathcal{A}|E_n) = \sigma(\mathcal{A})|E_n$, (Verify this!) and that $\widetilde{\mu}_i|_{\mathcal{A}|E_n}, i = 1, 2$ yield extensions of $\mu|_{\mathcal{A}|E_n}$ to $\sigma(\mathcal{A})|E_n$; deduce from the already proved finite case that $\widetilde{\mu}_1(A) = \widetilde{\mu}_2(A)$ if $A \in \sigma(\mathcal{A}|E_n)$ for some n, and hence that $\widetilde{\mu}_1 \equiv \widetilde{\mu}_2$.   □

Having got the extension theorem out of the way, we shall henceforth always assume that a measure is defined on a σ-algebra of sets. We make a formal definition.

Definition 1.5.12

(1) *By* **measurable space**, *we shall mean a pair* $(\Omega, \mathcal{B})$, *where* $\mathcal{B}$ *is a* σ-*algebra of subsets of a set* Ω.

(2) *A* **measure space** *is a triple* $(\Omega, \mathcal{B}, \mu)$ *where* μ *is a measure defined on a* σ-*algebra* $\mathcal{B}$ *of subsets of a set* Ω.

(3) *A measure space* $(\Omega, \mathcal{B}, \mu)$ *is said to be a* **finite** *(resp.,* σ-**finite,** *or* **probability***) measure space according as* μ *is a finite (resp.,* σ-*finite, or probability) measure.*

EXERCISES

Ex. 1.5.13   Let $\mathcal{A}$ denote the algebra $\mathcal{A} = \mathcal{A}(\mathcal{S})$ considered in Ex. 1.3.7. Let $\Omega = \mathbb{Q}$ denote the set of rational numbers in $\mathbb{R}$, and let $\mathcal{A}_0 = \mathcal{A} \cap \mathbb{Q}$. Consider the measure μ defined on $\mathcal{A}_0$ by $\mu(A) = 0$ or ∞ according to whether $A = \emptyset$ or $A \neq \emptyset$. Show that there exist more than one measure on $\sigma(\mathcal{A}_0)$ (in $\mathbb{Q}$) which agree with μ on $\mathcal{A}_0$. (Hence, the extension guaranteed by Theorem 1.5.8 is not always unique.)

Ex. 1.5.14   Suppose $(\Omega, \mathcal{B}, \mu)$ is a finite measure space, and suppose $\mathcal{A}$ is an algebra of subsets of Ω such that $\mathcal{B} = \sigma(\mathcal{A})$. Show that if $B \in \mathcal{B}$, and if $\varepsilon > 0$, then there exists $A \in \mathcal{A}$ such that $\mu(A \Delta B) < \varepsilon$, where $A \Delta B = (A - B) \cup (B - A)$. (*Hint:* Show that the collection of sets B in $\mathcal{B}$ for which the desired conclusion holds is a monotone class containing $\mathcal{A}$.)

Ex. 1.5.15   Let $(\Omega, \mathcal{B}, \mu)$ be a σ-finite measure space, let $\mathcal{A}$ be any algebra of subsets of Ω such that $\mathcal{B} = \sigma(\mathcal{A})$, and let $\mu_0 = \mu|\mathcal{A}$.

(1)   Show that $\mu_0^* = \mu^*$ (as functions on 2^{Ω}) and that, in fact,

$$\mu_0^*(E) = \mu^*(E) = \inf\{\mu(B) : B \in \mathcal{B}, E \subseteq B\}$$

for all subsets E of Ω.

(2)   If $N \subseteq \Omega$ and $\mu^*(N) = 0$, show that $N \in \mathcal{M}(\mu)$; we say that $(\Omega, \mathcal{M}(\mu), \mu^*)$ is a complete measure space, meaning that if $N \subseteq M, M \in \mathcal{M}(\mu)$ and $\mu^*(M) = 0$, then $N \in \mathcal{M}(\mu)$, i.e., $\mathcal{M}(\mu)$ contains all μ^*-null sets.

(3)   Show that $E \in \mathcal{M}(\mu)$ if and only if there exist $B_0, B_1 \in \mathcal{B}$ such that $B_0 \subseteq E \subseteq B_1$ and $\mu(B_1 - B_0) = 0$. (*Hint:* First assume $\mu(\Omega) < \infty$; the general case easily follows from this by the assumed σ-finiteness. Use (1) to lay hands on B_1. Define B_0 to be the complement of the B_1 you would have got for E'.)

(4) Make precise the statement, '$(\Omega, \mathcal{M}(\mu), \mu^*)$ is the completion of $(\Omega, \mathcal{B}, \mu)$'.

1.6 Lebesgue Measure

Assume throughout this section that:

(1) $\Omega = \mathbb{R}$.
(2) $\mathcal{I} = \{(a, b] : -\infty \leq a < b \leq \infty\}$—with the understanding that $(a, \infty] = (a, \infty)$.
(3) $\mathcal{A} = \mathcal{A}(\mathcal{I}) = \{\bigsqcup_{i=1}^{n} A_i : A_i \in \mathcal{I}, n = 0, 1, 2, \ldots\}$ (cf. Ex. 1.3.7).
(4) $\mathcal{B} = \sigma(\mathcal{A})$, the Borel σ-algebra on $\mathbb{R}$.

The goal of this chapter is to prove the following theorem.

Theorem 1.6.1   *There exists a unique measure μ defined on $\mathcal{B}$ such that $\mu((a, b]) = b - a$ for $a < b$. This measure, called the **Lebesgue measure** on $\mathbb{R}$, is **translation invariant**, i.e., $\mu(B + x) = \mu(B)$ for all $B \in \mathcal{B}$ and $x \in \mathbb{R}$.*

The proof of this theorem will be via the Caratheodory extension theorem. The thing to be verified is that the obvious definition of μ on $\mathcal{A}$ is well-defined and countably additive. This depends on a 'compactness argument'. We will need the fact that if a closed bounded interval $[c, d]$ is contained in an infinite union $\bigcup_n (c_n, d_n)$ of open intervals, then there is an integer N such that $[c, d] \subseteq \bigcup_{n=1}^{N} (c_n, d_n)$. We do the necessary verification through a couple of lemmas.

Lemma 1.6.2   *If $\bigsqcup_n (a_n, b_n] \subseteq (a, b]$, where $\{(a_n, b_n]\}_n$ is a finite or countable collection of intervals, then*

$$\sum_n (b_n - a_n) \leq b - a.$$

Proof: Suppose first that there are only finitely many, say m, non-empty intervals. Assume, without loss of generality, that the labelling is such that $a_1 < a_2 < \ldots < a_m$. The assumed disjointness then implies that

$$a \leq a_1 < b_1 \leq a_2 < b_2 \leq \ldots \leq a_m < b_m \leq b,$$

from which it is clear that

$$b - a = (b - b_m) + \sum_{k=2}^{m} (a_k - b_{k-1}) + (a_1 - a) + \sum_{k=1}^{m} (b_k - a_k)$$

$$\geq \sum_{k=1}^{m} (b_k - a_k).$$

If there are infinitely many intervals, the already established finite case shows that

$$\sum_{k=1}^{m}(b_k - a_k) \le b - a$$

for each m, and hence also

$$\sum_{k=1}^{\infty}(b_k - a_k) \le b - a. \qquad \square$$

Lemma 1.6.3   *If $(a, b] \subseteq \cup_n(a_n, b_n]$, where $\{(a_n, b_n]\}_n$ is a finite or countably infinite collection of intervals, then*

$$b - a \le \sum_n(b_n - a_n).$$

Proof: The case of finitely many intervals is easy and left as an exercise to the reader.

Suppose then that $(a, b] \subseteq \cup_{n=1}^{\infty}(a_n, b_n]$. Fix $\varepsilon > 0$, and note that

$$[a + \varepsilon, b] \subseteq \cup_{n=1}^{\infty}\left(a_n, b_n + \frac{\varepsilon}{2^n}\right).$$

By compactness, we can find an integer N such that

$$[a + \varepsilon, b] \subseteq \cup_{n=1}^{N}\left(a_n, b_n + \frac{\varepsilon}{2^n}\right).$$

It follows from the finite case that

$$b - a - \varepsilon \le \sum_{n=1}^{N}\left(b_n + \frac{\varepsilon}{2^n} - a_n\right).$$

Whence,

$$b - a \le \sum_{n=1}^{\infty}(b_n - a_n) + 2\varepsilon. \qquad \square$$

It follows from Lemma 1.6.2 and Lemma 1.6.3 that the equation

$$\mu\left(\coprod_{i=1}^{n}(a_i, b_i]\right) = \sum_{i=1}^{n}(b_i - a_i)$$

yields an unambiguously defined set function μ on the algebra $\mathcal{A}$ which is countably additive. That is, μ is a measure on $\mathcal{A}$. Clearly μ is σ-finite.

It follows from Theorem 1.5.11 that μ extends uniquely to a measure defined on $\mathcal{B} = \sigma(\mathcal{A})$. Since $\mu(A + x) = \mu(A)$ for all $A \in \mathcal{A}$ and $x \in \mathbb{R}$, it follows from the uniqueness of the extension that we have $\mu(B + x) = \mu(B)$ for all $B \in \mathcal{B}$ and $x \in \mathbb{R}$. This completes the proof of Theorem 1.6.1.

We conclude the section with an exercise that simultaneously establishes the existence of at least one set $E \subseteq \mathbb{R}$ which is not Lebesgue measurable—i.e., $E \notin \mathcal{M}(\mu)$—and the fact that there exists no measure defined on all subsets of $\mathbb{R}$ which is translation invariant and assigns measure 1 to the unit interval $[0,1]$.

EXERCISES

Ex. 1.6.4

(1) Let $\mathcal{L} = \mathcal{M}(\mu)$ denote the σ-algebra of μ^*-measurable sets (cf. Theorem 1.5.8) where μ denotes the Lebesgue measure. (Members of $\mathcal{L}$ are called **Lebesgue-measurable** sets.) Then, show that: $L \in \mathcal{L}, x \in \mathbb{R} \Rightarrow L + x \in \mathcal{L}$ and that $\mu^*(L + x) = \mu^*(L)$. (Hence μ^* is a translation-invariant measure defined on $\mathcal{L}$.)

(2) Let $I = (0, 1]$, and for $x, y \in I$, say $x \sim y$ if $x - y$ is a rational number. For $x, y \in (0, 1]$, define $x \overset{\circ}{+} y$ to be the unique element of $(0,1]$ such that $(x + y) - (x \overset{\circ}{+} y)$ is an integer.

    (a) Show that '$\sim$' is an equivalence relation on I, and that if $x, y \in (0, 1]$, then $x \sim y$ if and only if there exists a rational number $r \in (0, 1]$ such that $x = y \overset{\circ}{+} r$.

    (b) If $[x]$ denotes the equivalence class of an element x of I so that $[x] = \{y \in I : x \sim y\}$, let E be a set obtained by picking one element from each distinct equivalence class. Let $\{r_n\}_{n=1}^{\infty}$ be an enumeration of the rational numbers in $(0, 1]$, and define

$$E_n = E \overset{\circ}{+} r_n = \{x \overset{\circ}{+} r_n : x \in E\}.$$

    Show that $(0, 1] = \coprod_{n=1}^{\infty} E_n$; also show that if $E \in \mathcal{L}$, then $E_n \in \mathcal{L}$ and $\mu^*(E_n) = \mu^*(E)$ for all n; deduce that $E \notin \mathcal{L}$.

Ex. 1.6.5
Let $I = [0, 1]$. Let $I_1 = I_{11} = (\frac{1}{3}, \frac{2}{3})$ be the open middle-third interval of I. Next, let $I_{21} = (\frac{1}{9}, \frac{2}{9})$ and $I_{22} = (\frac{7}{9}, \frac{8}{9})$ be the two open middle-third intervals of $I - I_1$. Let $I_2 = I_{21} \cup I_{22}$. For $j \geq 3$ and $k = 1, 2, 3, \ldots, 2^{j-1}$, let I_{jk} be the open middle-third intervals of $I - \cup_{k=1}^{j-1} I_k$ and let $I_j = \cup_{k=1}^{2^{j-1}} I_{jk}$. Finally, let $C = I - \cup_{j=1}^{\infty} I_j$; C is called the **Cantor set** .

(1) Show that C is compact and uncountable.
(2) Show that $\lambda(C) = 0$, where λ is the Lebesgue measure on $[0, 1]$.

Ex. 1.6.6

(1) If μ is a probability measure defined on the Borel σ-algebra $\mathcal{B}$ of $\mathbb{R}$, define $F : \mathbb{R} \to [0, 1]$ by $F(x) = \mu((-\infty, x])$ and verify that

   (a) F is monotonically non-decreasing, i.e. $x \leq y \Rightarrow F(x) \leq F(y)$, and right-continuous, i.e., $\lim_{y \downarrow x} F(y) = F(x)$;
   (b) F is discontinuous at x if and only if $\mu(\{x\}) > 0$ and
   (c) $\lim_{x \to \infty} F(x) = 1$, $\lim_{x \to -\infty} F(x) = 0$.
       The function F is referred to as the **distribution function** of μ.

(2) Conversely, if $F : \mathbb{R} \to [0, 1]$ is a function satisfying (a) and (c) above, (imitate the construction of the Lebesgue measure to) show that there exists a unique probability measure μ on $\mathbb{R}$ such that $\mu((-\infty, x]) = F(x)$ for all x in $\mathbb{R}$.

(3) Generalise (1) and (2) above to the case of σ-finite (rather than just probability) measures.

1.7   Regularity

This small section is devoted to a brief encounter with **regularity**. We begin with a few simple exercises.

Ex. 1.7.1 Let (X, d) be a metric space. For a set $A \subset X$, define the function $d_A : X \to [0, \infty)$ by

$$d_A(x) \ = \ \text{dist}(x, A) \ = \ \inf\{d(x, a) : a \in A\}.$$

Show that

(1) $|d_A(x) - d_A(y)| \leq d(x, y)$ for any $x, y \in X$;
(2) d_A is a continuous function, and

$$\{x \in X : d_A(x) = 0\} = \bar{A};$$

(3) if A is a closed set in a metric space, there exists a continuous function $f : X \to \mathbb{R}$ such that $A = f^{-1}(\{0\})$;
(4) every closed set in a metric space is a G_δ set, i.e., is expressible as an intersection of countably many sets. (*Hint:* In the notation of (3) above, let $U_n = \{x \in X : f(x) < \frac{1}{n}\}$.);
(5) if A and B are disjoint closed sets in X, then there exists a continuous function $f : X \to [0, 1]$ such that $A = f^{-1}(\{0\})$ and $B = f^{-1}(\{1\})$. (*Hint:* Consider $f(x) = \frac{d_A(x)}{d_A(x) + d_B(x)}$.)

Proposition 1.7.2   *Suppose μ is a finite positive measure defined on the Borel σ-algebra $\mathcal{B}_X$ of a compact metric space. Then, for any Borel set $E \in \mathcal{B}_X$, we have*

$$\mu(E) = \sup\{\mu(K) : K \subset E \text{ and } K \text{ is compact}\}$$
$$= \inf\{\mu(U) : E \subset U \text{ and } U \text{ is open}\}.$$

Proof: Let $\mathcal{M}$ denote the class of all those $E \in \mathcal{B}_X$ which satisfy the conclusion of Proposition 1.7.2.

Next, suppose that K is a compact set in X. Then the first of the desired identities is clearly satisfied by K. On the other hand, we may conclude from Ex. 1.7.1(4) that there exists a sequence $\{V_n\}_{n=1}^{\infty}$ of open sets such that $K = \cap_{n=1}^{\infty} V_n$. It is then seen from the fact—see Proposition 1.4.6(2)—that finite measures are 'continuous from above' that if we set $U_n = \cap_{m=1}^{n} V_m$, then

$$\mu(K) = \lim_{n \to \infty} \mu(U_n);$$

hence, K also satisfies the second of the desired identities. Thus $\mathcal{M}$ contains all compact sets.

We claim next that $\mathcal{M}$ is an algebra of sets. So, suppose $E_i \in \mathcal{M}$ for $i = 1, 2$. Let $\epsilon > 0$. Then we can find compact sets K_1, K_2 and open sets U_1, U_2 such that $K_i \subset E_i \subset U_i$ and $\mu(U_i \setminus K_i) < \epsilon$. Then, clearly, $K = K_1 \cup K_2$ is a compact subset of $E = E_1 \cup E_2$ and $U = U_1 \cup U_2$ is an open superset of E and we have

$$\mu(U \setminus K) \leq \mu(U_1 \setminus K) + \mu(U_2 \setminus K)$$
$$\leq \mu(U_1 \setminus K_1) + \mu(U_2 \setminus K_2)$$
$$< 2\epsilon;$$

the arbitrariness of ϵ shows that indeed $E \in \mathcal{M}$.

Since complements of compact sets in X are open (and conversely), the class $\mathcal{M}$ is seen to be closed under the formation of complements. Hence, $\mathcal{M}$ indeed contains the algebra generated by all compact sets in X. Hence, in order to complete the proof, it suffices, by Proposition 1.3.2, to verify that $\mathcal{M}$ is a monotone class of sets.

So suppose $E_n \in \mathcal{M}$, $n = 1, 2, \ldots$, and let $\epsilon > 0$. Pick compact sets K_n and open sets U_n such that $K_n \subset E_n \subset U_n$ and $\mu(U_n \setminus K_n) < \frac{\epsilon}{2^n}$ for each n.

If $E_n \uparrow E$, pick a large m so that $\mu(E \setminus E_m) < \epsilon$, and conclude, with $U = \cup_{n=1}^{\infty} U_n$ and $K = K_m$, that U is open, K is compact, $K \subset E \subset U$ and that

$$\mu(U \setminus E) \leq \sum_{n=1}^{\infty} \mu(U_n \setminus E)$$
$$\leq \sum_{n=1}^{\infty} \mu(U_n \setminus E_n)$$

$$\leq \sum_{n=1}^{\infty} \mu(U_n \setminus K_n)$$

$$< \epsilon,$$

while

$$\mu(E \setminus K_m) \leq \mu(E \setminus E_m) + \mu(E_m \setminus K_m) < 2\epsilon.$$

Since ϵ was arbitrary, this shows that $E \in \mathcal{M}$.

The case when $E_n \downarrow E$ is similarly seen (take $U = U_k$ and $K = \cap_n K_n$, with k so large that $\mu(E_k \setminus E) < \epsilon$) to imply that $E \in \mathcal{M}$, thereby showing that $\mathcal{M}$ is a monotone class containing the algebra generated by the collection of all compact sets. This completes the proof of the proposition. □

The above proposition is usually rephrased thus: 'every finite positive measure on a compact metric space is **regular**'. (The two equations in the proposition are referred to as equations of **inner regularity** and **outer regularity**, respectively.) The reader interested in such regularity properties of measures on more general spaces (compact non-metrizable, non-compact, locally compact spaces, etc.) may consult [Hal], for instance.

EXERCISES

Ex. 1.7.3   If P is a probability measure defined on $(\mathbb{R}, \mathcal{B}_{\mathbb{R}})$, and if $E \in \mathcal{B}_{\mathbb{R}}$, then for all $\varepsilon > 0$, prove that there exists an open set U and a compact set $K \subseteq E \subseteq U$ and $P(U \setminus K) < \varepsilon$. *(Hint: See Lemma 4.4.1.)*

1.8   Bernoulli Trials

Let us now revisit the experiment (Experiment 4 in Section 1.1) of 'picking a point at random from (0,1]'. A natural choice is to take $\Omega = (0, 1]$, $\mathcal{B}$ to the Borel-σ algebra, and P to be the Lebesgue measure on $(0, 1]$. Here,

$$P(\text{the point chosen lies in } (a, b) \subset (0, 1]) \equiv P((a, b)) = b - a,$$

hence, proportional to the length of (a, b) in $(0, 1]$, thus making the case that every point is chosen in an 'equally likely' manner.

We wish to discuss another interpretation of this experiment in this section.

Bernoulli trials: Toss an unbiased coin (i.e. head and tails are equally likely to result) infinitely many times, and suppose the result of each toss has no effect on any other toss. If an outcome is head, record it as 1, otherwise 0. Clearly, the outcome space is $\Omega = \{0, 1\}^{\mathbb{N}}$.

Before we set up σ-algebras and P, we establish the connection with Experiment 4. We refer the reader to [3] for details. The key ingredient in making the connection is the fact that each x in $(0,1]$ admits a dyadic (= base 2) expansion $x = \sum_{n=1}^{\infty} a_n/2^n$, with $a_n \in \{0, 1\}$, and that such an expansion is unique if we decide to consistently use the non-terminating expansion in case x is a dyadic rational number (i.e., a number of the form $k/2^n$, where k, n are integers). (Thus, e.g., if $x = 1/2$, we write $x = 1/2^2 + 1/2^3 + 1/2^4 + \ldots$ rather than $x = 1/2 + 0/2^2 + 0/2^3 + \ldots$). Hence an outcome of the Bernoulli trials (= infinite coin-tossing experiment) can be translated to an outcome of choosing a random real number from $(0, 1]$. Consequently, a subset E of $(0,1]$ can be thought of as the event that the randomly selected point of $(0,1]$ lies in E, or, equivalently, as the event that the infinite sequence $\{a_n\}_{n=1}^{\infty}$ of outcomes of a tossed coin satisfies $\sum \frac{a_n}{2^n} \in E$.

Now, we get back to $\Omega = \{0, 1\}^{\mathbb{N}}$. Let $\pi_n : \Omega \to \Omega_n$ denote the projection on to the first n coordinates, where $\Omega_n = \{(w_1, \ldots, w_n) : w_i = 0 \text{ or } 1\}$ is the natural sample space describing n successive coin tosses. If $E \subseteq \Omega_n$, then $\pi_n^{-1}(E)$ is the typical n-dimensional cylinder set (with base E) in Ω, and the collection $\mathcal{A}_n = \{\pi_n^{-1}(E) : E \subseteq \Omega_n\}$ of all such n-dimensional cylinders is a finite algebra of subsets of Ω. (This is the algebra consisting of those events which are determined by the first n tosses. All this is discussed in some detail and in much greater generality in Section 4.3. Since $\mathcal{A}_n \subseteq \mathcal{A}_{n+1}$ $\forall n$, it is clear that $\mathcal{A} = \cup_{n=1}^{\infty} \mathcal{A}_n$ is an algebra of subsets of Ω.

On the other hand, the space $\Omega = \{0, 1\}^{\mathbb{N}}$ is compact in the so-called 'product topology' of the discrete spaces $\{0, 1\}$; it is a metric space with respect to $d(w, w') = \sum_{n=1}^{\infty} 2^{-n}|w_n - w_n'|$. Since every subset of Ω_n is closed (in the discrete space Ω_n), it follows that if $A \in \mathcal{A} = \cup_{n=1}^{\infty} \mathcal{A}_n$, then A is a compact subset of Ω. Since this is also true of A', we see that each $A \in \mathcal{A}$ is a compact open set in Ω. In fact, the algebra $\mathcal{A}$ forms a base for the topology on Ω, and (since Ω is compact) it follows that $\sigma(\mathcal{A}) = \mathcal{B}_{\Omega}$, the natural Borel σ-algebra of Ω (cf. Ex. 1.2.12). Let us write $\mathcal{B} = \sigma(\mathcal{A})$. We conclude this section with two problems that indicate the above-mentioned relationship.

EXERCISES

Ex. 1.8.1 Let $\Omega, \mathcal{B}_n, \mathcal{A}, \mathcal{B}$ be as above. Let μ be a probability measure defined on $\mathcal{B}$. Define $\mu_n = \mu|\mathcal{B}_n$. Then μ_n is a probability measure defined on $\mathcal{B}_n$, and the $\mu_n's$ satisfy the consistency condition $\mu_n = \mu_{n+1}|\mathcal{B}_n$.

Conversely, if μ_n is a probability measure defined on $\mathcal{B}_n$, and if $\mu_{n+1}|\mathcal{B}_n = \mu_n$, then we may define the set function μ on $\mathcal{A} = \cup_{n=1}^{\infty} \mathcal{B}_n$ by $\mu|\mathcal{B}_n = \mu_n$. Verify that μ is finitely additive on $\mathcal{A}$, deduce that μ is a well-defined countably additive set function, and conclude that μ extends uniquely to a probability measure defined on $\sigma(\mathcal{A}) = \mathcal{B}$. (*Hint:* If a compact set is the union of countably many pairwise disjoint open sets, then all but finitely many of those open sets must be empty.)

As a special case of the above exercise, define $\mu_n : \mathcal{B}_n \to [0, 1]$ by $\mu_n\left(\pi_n^{-1}(E)\right) = |E|.2^{-n}$, where $|E|$ denotes the cardinality of the subset E of Ω_n. The resulting measure, μ, on $\mathcal{B}$ is called the measure underlying a sequence of Bernoulli trials (with a fair coin).

Ex. 1.8.2   In what sense can the function $T : \Omega \to [0, 1]$ defined by $T(w) = \sum_{n=1}^{\infty} w_n 2^{-n}$ be regarded as an an 'isomorphism between the measure spaces' $(\Omega, \mathcal{B}, \mu_{\text{Bernoulli}})$ and $([0, 1], \mathcal{B}_{[0,1]}, \mu_{\text{Lebesgue}})$?

Ex. 1.8.3   Let $\pi : \Omega \to \Omega'$ be any function and $\mathcal{B}' \subset 2^{\Omega'}$ any σ-algebra. Define $\mathcal{B} = \{\pi^{-1}(E') : E' \in \mathcal{B}'\}$. Show that $\mathcal{B}$ is a σ-algebra of subsets of Ω and that $\mu \to \mu \circ \pi^{-1}$ defines a bijection from the set of measures on $(\Omega, \mathcal{B})$ to $(\Omega', \mathcal{B}')$.

2

Integration

2.1 Measurable Functions

This chapter is devoted to basic integration theory. Thus, for instance, if f is a suitable real-valued function $f : \mathbb{R} \to \mathbb{R}$, we would like to make sense of $\int f d\mu = \int f(x) d\mu(x)$, where μ denotes Lebesgue measure; this notion should, to be of any use, agree intuitively with 'the area under the graph' (if $f \geq 0$). In the simplest case where $f = 1_E$ (thus $f(x) = 0$ or 1 according as $x \notin E$ or $x \in E$), we would like the value of $\int 1_E d\mu$ to be $\mu(E)$; however $\mu(E)$ makes sense only for E in $\mathcal{B}$ (or at most $\mathcal{L} = \mathcal{M}(\mu)$), and thus, we should not expect to be able to integrate all functions, but only the 'measurable' ones.

For convenience of exposition, we define the notion of measurability in greater generality (and will soon realise the advantages of the generality).

Definition 2.1.1 *Let $(\Omega, \mathcal{B})$ and $(\Omega', \mathcal{B}')$ be measurable spaces, cf. Definition 1.5.12 (1). A function $T : \Omega \to \Omega'$ is said to be* **measurable** *(or $(\mathcal{B}, \mathcal{B}')$-measurable if we wish to stress the dependence on the σ-algebras $\mathcal{B}$ and $\mathcal{B}'$) if $T^{-1}(B') \in \mathcal{B}$ for all $B' \in \mathcal{B}'$.*

A function $T : \Omega \to \mathbb{R}$ is said to be measurable if T is $(\mathcal{B}, \mathcal{B}_\mathbb{R})$-measurable, where $\mathcal{B}_\mathbb{R}$ is the Borel σ-algebra of $\mathbb{R}$.

The proof of the following proposition is elementary and is left an exercise to the reader.

Proposition 2.1.2

(1) *If $(\Omega_i, \mathcal{B}_i), i = 1, 2, 3$ are measurable spaces, if $T : \Omega_1 \to \Omega_2$ is $(\mathcal{B}_1, \mathcal{B}_2)$-measurable and $S : \Omega_2 \to \Omega_3$ is $(\mathcal{B}_2, \mathcal{B}_3)$-measurable, then $S \circ T : \Omega_1 \to \Omega_3$ is $(\mathcal{B}_1, \mathcal{B}_3)$-measurable.*

(2)   *If* $(\Omega, \mathcal{B})$ *and* $(\Omega', \mathcal{B}')$ *are measurable spaces, and if* $\mathcal{S}'$ *is any collection of subsets of* Ω' *such that* $\mathcal{B}' = \sigma(\mathcal{S}')$, *then* T *is* $(\mathcal{B}, \mathcal{B}')$-*measurable if and only if* $T^{-1}(E) \in \mathcal{B}$ *for all* $E \in \mathcal{S}'$. *In particular, a function* $f : \Omega \to \mathbb{R}$ *is measurable (i.e.,* $(\mathcal{B}, \mathcal{B}_\mathbb{R})$-*measurable) if and only if* $f^{-1}(E) \in \mathcal{B}$ *for all* $E \in \mathcal{S}$, *where* $\mathcal{S}$ *is any one of the collections* $\mathcal{S}_1$, $\mathcal{S}_2$, $\mathcal{S}_3$, $\mathcal{U}$, $\mathcal{C}$ *or* $\mathcal{K}$ *defined in Ex. 1.2.13.*

We now proceed to see that there are many measurable functions, and to essentially describe what they all look like. We start with an elementary observation, stated as a lemma, whose obvious proof we omit.

Lemma 2.1.3   *The following conditions on a function* $f : \Omega \to \mathbb{R}$, *whose range* $f(\Omega)$ *is a finite subset of* $\mathbb{R}$, *are equivalent:*

(1) f *is* $(\mathcal{B}, \mathcal{B}_\mathbb{R})$-*measurable.*
(2)   *There exist* $E_i \in \mathcal{B}$ *and* $a_i \in \mathcal{B}$ *such that* $f = \sum_{i=1}^{n} a_i 1_{E_i}$.
(3)   *There exist* a_i *and* E_i *as above, with the additional property that* $\Omega = \coprod_{i=1}^{n} E_i$.

A function f as in the above lemma is said to be a **simple function**. If, in (2) above, it is demanded in addition that $f(\Omega) = \{a_1, \ldots, a_n\}$ and $E_i = f^{(-1)}(\{a_i\})$ $\forall i$, then the a_i's and E_i's are uniquely determined by f. (We shall sometimes refer to $f = \sum_{i=1}^{n} a_i 1_{E_i}$ as the canonical decomposition of the simple function f.)

We next establish a fundamental approximation theorem, which we shall repeatedly have to apply, which describes what every non-negative measurable function looks like.

Proposition 2.1.4   *A non-negative measurable function is the pointwise limit of an increasing sequence of non-negative simple functions; i.e., there exist simple functions* $\{f_n\}_{n=1}^{\infty}$ *such that* $0 \leq f_1(x) \leq f_2(x) \leq \ldots \nearrow f(x)$.

Proof: In a sense, the proof of this proposition lays bare the fundamental idea of integration, wherein the area under the graph of a function is approximated by a finite union of rectangles from within. The idea behind the proof is the following: if $f : \Omega \to \mathbb{R}$ is a non-negative measurable function, and if n is any positive integer, let $\{I_{n,k}\}_{k=1}^{\infty}$ be the partition of $[0, \infty)$ into a 'grid' of size 2^{-n}; in other words, let $I_{n,k} = \left[\frac{k}{2^n}, \frac{k+1}{2^n}\right)$, $E_{n,k} = f^{-1}(I_{n,k})$, and define $h_n = \sum_{k=0}^{\infty} \frac{k}{2^n} 1_{E_{n,k}}$; it must be clear that the h_n's inherit measurability from f, that $0 \leq h_1 \leq h_2 \leq \ldots$, and that $h_n \leq f < h_n + 1/2^n$ so that, in particular, the h_n's converge uniformly to f.

The h_n's are 'elementary' functions, i.e., they have countable range, although they are not simple unless f is bounded. The proposition is proved by setting $f_n = h_n \wedge n = \min\{h_n, n\}$. $\square$

Remark 2.1.5 *It must be clear from the proof of Proposition 2.1.4 that any bounded measurable function is a uniform limit of simple functions.*

EXERCISES

Ex. 2.1.6 If $T : X \to X'$ is a continuous map between topological spaces, then T is $(\mathcal{B}_X, \mathcal{B}_{X'})$-measurable, where we write $\mathcal{B}_X$ to denote the σ-algebra generated by all open sets in X.

Ex. 2.1.7

(1)   If $\mathcal{B}_i$ is a σ-algebra of subsets of a set X_i, and if $f_i : X \to X_i$ are functions, for each $i \in I$, where I is some index set, show that there is a smallest σ-algebra $\mathcal{B}$ of subsets of X such that f_i is $(\mathcal{B}, \mathcal{B}_i)$-measurable for each $i \in I$. If we denote this σ-algebra $\mathcal{B}$ by $\sigma(\{f_i : i \in I\})$, show that

$$\sigma(\{f_i : i \in I\}) = \sigma(\{f_i^{(-1)}(U_i) : U_i \in \mathcal{S}_i, i \in I\}),$$

for any family $\{\mathcal{S}_i : i \in I\}$ such that $\sigma(\mathcal{S}_i) = \mathcal{B}_i \ \forall i$.

(2)   If $X = \prod_{i \in I} X_i$, where $(X_i, \mathcal{U}_i)$ are topological spaces, and if X is endowed with the product topology $\mathcal{U}$, verify that $\mathcal{B}_X = \sigma(\{\pi_i : i \in I\})$, where $\pi_i : X \to X_i$ is the natural projection map.

Ex. 2.1.8

(1)   Let $f : \Omega \to \mathbb{R}^n$ be a function. Then there exist unique functions $f_i : \Omega \to \mathbb{R}$, $1 \le i \le n$, such that $f(w) = (f_1(w), f_2(w), \dots, f_n(w)) \forall w \in \Omega$.

(2)   If $\mathcal{S} = \{\prod_{i=1}^{n}(a_i, b_i) : -\infty < a_i < b_i < \infty \ \forall i\}$ is the set of open rectangular boxes in $\mathbb{R}^n$, show that $\mathcal{B}_{\mathbb{R}^n} = \sigma(\mathcal{S})$.

(3)   If $(\Omega, \mathcal{B})$ is a measurable space, and if $f, f_1, \dots, f_n$ are as in (1) above, show that f is $(\mathcal{B}, \mathcal{B}_{\mathbb{R}^n})$-measurable if and only if f_i is $(\mathcal{B}, \mathcal{B}_{\mathbb{R}})$-measurable for each $i = 1, 2, \dots, n$.

(4)   If $f_1, \dots, f_n : \Omega \to \mathbb{R}$ are measurable functions and if $\phi : \mathbb{R}^n \to \mathbb{R}$ is continuous (in fact, measurability of ϕ is enough), then $\phi \circ (f_1, \dots, f_n)$ (i.e., the function $g : \Omega \to \mathbb{R}$ defined by $g(w) = \phi((f_1(w), \dots, f_n(w)))$ is measurable).

In particular, if $f, f_1, \dots, f_n : \Omega \to \mathbb{R}$ are measurable, then each of the following functions (from Ω to $\mathbb{R}$) is measurable:

(a) $\sum_{i=1}^{n} a_i f_i$, where $a_1, \dots, a_n$ are arbitrary real numbers;

(b) $\prod_{i=1}^{n} f_i$;

(c) $|f|$;

(d) $f_{\pm} = \frac{|f| \pm f}{2}$;

(e) $f_1 \vee f_2 = \max\{f_1, f_2\} = \frac{f_1 + f_2 + |f_1 - f_2|}{2}$;

(f) $f_1 \wedge f_2 = \min\{f_1, f_2\} = \frac{f_1 + f_2 - |f_1 - f_2|}{2}$.

Ex. 2.1.9

(1)   Make precise the notion of measurability for an extended real-valued function, i.e., a function which takes values in $\bar{\mathbb{R}} = \mathbb{R} \cup \{\pm\infty\}$.
(2)   What should be the natural σ-algebra of sets in the space $\mathbb{R}^\infty$ of real sequences? (Is there a 'natural' metric in $\mathbb{R}^\infty$?)

2.2   Integration

Let $(\Omega, \mathcal{B}, \mu)$ denote a measure space, fixed once and for all. Our first step in this section is to prove the following proposition, where we use the notation $\mathcal{S}_+$ to denote the class of non-negative simple functions defined on $(\Omega, \mathcal{B}, \mu)$.

Proposition 2.2.1   *There exists a unique 'functional'*

$$\mathcal{S}_+ \ni f \mapsto \int f d\mu \in [0, \infty]$$

which satisfies the two conditions below:

(1) $\int 1_E d\mu = \mu(E) \; \forall \, E \in \mathcal{B}$
(2) $\int (af + bg) d\mu = a \int f d\mu + b \int g d\mu \; \forall f, g \in \mathcal{S}_+, a, b > 0$, *where* $a \cdot \infty$ *is understood as* ∞.

Proof: We start with the following:

Assertion: If $\Omega = \coprod_{i=1}^m A_i = \coprod_{j=1}^n B_j$, then

$$\sum_{i=1}^m a_i 1_{A_i} = \sum_{j=1}^n b_j 1_{B_j} \; \Rightarrow \; \sum_{i=1}^m a_i \mu(A_i) = \sum_{j=1}^n b_j \mu(B_j) \,.$$

This assertion is true because

$$A_i \cap B_j \neq \emptyset \Rightarrow a_i = b_j$$

and hence, by additivity of measures,

$$\sum_{i=1}^m a_i \mu(A_i) = \sum_{i=1}^m a_i \mu \left(\coprod_{j=1}^n A_i \cap B_j \right)$$

$$= \sum_{i=1}^{m} \sum_{j=1}^{n} a_i \mu \left(A_i \cap B_j \right)$$

$$= \sum_{i=1}^{m} \sum_{j=1}^{n} b_j \mu \left(A_i \cap B_j \right)$$

$$= \sum_{j=1}^{n} b_j \mu \left(\coprod_{i=1}^{m} A_i \cap B_j \right)$$

$$= \sum_{j=1}^{n} b_j \mu(B_j) \, .$$

Now, if $f = \sum_{i=1}^{m} a_i 1_{A_i}$ is any function in $\mathcal{S}_+$, define

$$\int f d\mu = \sum_{i=1}^{m} a_i \mu(A_i).$$

The above assertion says that the value of $\int f d\mu$ is independent of the decomposition chosen. It is clear that this 'functional' satisfies condition (1) above. As for (2), if we have $f = \sum_{i=1}^{m} a_i 1_{A_i}$ and $g = \sum_{j=1}^{n} b_j 1_{B_j}$, where we assume (without loss of generality, thanks to the assertion) that $\Omega = \coprod_{i=1}^{m} A_i = \coprod_{j=1}^{n} B_j$, then, by two further applications of the above assertion, we have

$$\int (af + bg) d\mu = \int \left(a \sum_{i=1}^{m} a_i 1_{A_i} + b \sum_{j=1}^{n} b_j 1_{B_j} \right) d\mu$$

$$= \int \left(\sum_{i=1}^{m} \sum_{j=1}^{n} (aa_i + bb_j) 1_{A_i \cap B_j} \right) d\mu$$

$$= \sum_{i=1}^{m} \sum_{j=1}^{n} (aa_i + bb_j) \mu(A_i \cap B_j)$$

$$= \sum_{i=1}^{m} aa_i \mu \left(A_i \cap \coprod_{j=1}^{n} B_j \right) + \sum_{j=1}^{n} bb_j \mu \left(\coprod_{i=1}^{m} A_i \cap B_j \right)$$

$$= a \sum_{i=1}^{m} a_i \mu(A_i) + b \sum_{j=1}^{n} b_j \mu(B_j)$$

$$= a \int f d\mu + b \int g d\mu \, .$$

Finally, the uniqueness assertion is obvious. □

Corollary 2.2.2 $f, g \in \mathcal{S}_+$ and $f(\omega) \leq g(\omega) \ \forall \omega \in \Omega \Rightarrow \int f d\mu \leq \int g d\mu$.

Proof: In the special case when $f = 0$, this is true by definition of $\int g d\mu$; in general, if $h = g - f \in \mathcal{S}_+$, then by the special case discussed,

$$\int g d\mu = \int f d\mu + \int h d\mu \geq \int f d\mu .$$ □

We digress briefly with a discussion of limits and measurability.

In general, the class of measurable functions (or sets) is 'closed under count-able operations'. To illustrate this principle, for instance, if $f_n : \Omega \to \mathbb{R}$ is measurable for each n, let us verify that the set E of points at which the sequence converges to a finite limit is measurable; this is because

$$E = \{w : \{f_n(w)\}_{n=1}^{\infty} \text{ is convergent }\}$$

$$= \left\{ w : \ \forall \ k \exists N \ \forall m, n \geq N, |f_n(w) - f_m(w)| < \frac{1}{k} \right\}$$

$$= \cap_{k=1}^{\infty} \cup_{N=1}^{\infty} \cap_{m=N}^{\infty} \cap_{n=N}^{\infty} \left\{ w : |f_n(w) - f_m(w)| < \frac{1}{k} \right\} .$$

Similar manipulations show that if $\{f_n\}$ is a convergent sequences of real-valued functions, the limit function is measurable if each f_n is. More generally, $\liminf f_n$ and $\limsup f_n$ are measurable functions. Recall that

$$(\limsup f_n)(x) = \lim_{n \to \infty} \sup_{m \geq n} f_m(x)$$

$$= \inf_n \sup_{m \geq n} f_m(x),$$

and

$$(\liminf f_n)(x) = \lim_{n \to \infty} \inf_{m \geq n} f_m(x)$$

$$= \sup_n \inf_{m \geq n} f_m(x).$$

Also recall that if $\{a_n\}_{n=1}^{\infty}$ is any sequence of real numbers, than $\underline{a} = \liminf a_n = \lim_{n \to \infty} (\inf_{m \geq n} a_m)$ (resp., $\bar{a} = \limsup a_n = \lim_{n \to \infty} (\sup_{m \geq n} a_m)$) is the small-est (resp., largest) element of the set of possibly infinite limit points of all possible convergent subsequences of $\{a_n\}$; so that, in particular, a sequence $\{a_n\}_{n=1}^{\infty}$ converges (to a limit in $\bar{\mathbb{R}} = \mathbb{R} \cup \{\pm\infty\}$) if and only if $\underline{a} = \bar{a}$, in which case $\lim_{n \to \infty} a_n = \limsup a_n = \liminf a_n$.

Lemma 2.2.3 *Suppose $s_n, s \in S_+$ and $s_n \nearrow s$ (pointwise). Then*

$$\int s d\mu = \sup_n \int s_n d\mu = \lim_{n \to \infty} \int s_n d\mu \ .$$

Proof: On the one hand, by Corollary 2.2.2 we have

$$\int s d\mu \geq \sup_n \int s_n d\mu = \lim \sup_{n \to \infty} \int s_n d\mu. \qquad (2.2.1)$$

For the reverse inequality, we will need the following.

Assertion: If $t \in S_+$, then the function $v_t : B \to [0, \infty]$ defined by

$$v_t(E) = \int 1_E t d\mu,$$

(which makes sense since S_+ is closed under multiplication) is also a countably additive measure.

Reason: If $t = \sum_{i=1}^n a_i 1_{A_i}$, then

$$v_t \left(\coprod_{k=1}^\infty E_k \right) = \int \left(\sum_{i=1}^n a_i 1_{(\coprod_{k=1}^\infty E_k) \cap A_i} \right) d\mu$$

$$= \sum_{i=1}^n a_i \mu \left(\left(\coprod_{k=1}^\infty E_k \right) \cap A_i \right)$$

$$= \sum_{i=1}^n \sum_{k=1}^\infty a_i \mu (E_k \cap A_i)$$

$$= \sum_{k=1}^\infty v_t(E_k).$$

For obvious reasons, we shall write $\int_E t d\mu$ for $v_t(E)$.

Now, fix a constant $c \in (0, 1)$ and define $E_n = \{\omega : s_n(\omega) \geq cs(\omega)\}$; notice that $E_n \nearrow \Omega$ and conclude that

$$\int cs d\mu = \lim_{n \to \infty} \int 1_{E_n} cs d\mu$$

$$\leq \lim \inf_{n \to \infty} \int 1_{E_n} cs d\mu$$

$$\leq \lim \inf_{n \to \infty} \int s_n d\mu,$$

where we used the continuity from below the measure ν_{cs} in the first step, and the fact that $1_{E_n}cs \leq 1_{E_n}s_n$ (and that consequently $\int 1_{E_n}csd\mu \leq \int 1_{E_n}s_nd\mu$) in the last step.

Since this is valid for all $c < 1$, we find also that

$$\int sd\mu = \sup_{0<c<1} \int csd\mu \leq \liminf_{n\to\infty} \int s_nd\mu . \qquad (2.2.2)$$

The validity of the lemma is a consequence of (2.2.1) and (2.2.2). □

Lemma 2.2.4 *If $\{s_m : m = 1, 2, \ldots\}$ and $\{t_n : n = 1, 2, \ldots\}$ are two non-decreasing sequences in S_+ such that $\sup_m s_m(\omega) = \sup_n t_n(\omega) \,\forall \omega \in \Omega$, then*

$$\sup_m \int s_m d\mu = \sup_n \int t_n d\mu .$$

Proof: By two applications of Lemma 2.2.3, we see that

$$\sup_m \int s_m d\mu = \sup_m \left(\sup_n \int s_m \wedge t_n d\mu\right)$$

$$= \sup_{m,n} \int s_m \wedge t_n d\mu$$

$$= \sup_n \left(\sup_m \int s_m \wedge t_n d\mu\right)$$

$$= \sup_n \int t_n d\mu .$$

□

Proposition 2.1.4 and Lemma 2.2.4 now give us a clean recipe for unambiguously defining the integral of any non-negative measurable function f—choose any sequence $\{s_n\}$ of simple functions such that $0 \leq s_1(\omega) \leq s_2(\omega) \nearrow f(\omega) \,\forall \omega \in \Omega$ and define

$$\int fd\mu = \lim_{n\to\infty} \int s_n d\mu = \sup_n \int s_n d\mu .$$

We can now extend Proposition 2.2.1 to the case of arbitrary non-negative measurable (and not just simple) functions. First, let us agree to write $\mathcal{L}^0$ (resp., $\mathcal{L}^0_+$)   for the class of measurable (resp., non-negative measurable) functions.

Proposition 2.2.5   *There exists a unique 'functional'*

$$\mathcal{L}^0_+ \ni f \mapsto \int fd\mu \in [0, \infty]$$

which satisfies the following three conditions:

(1) $\int 1_E d\mu = \mu(E) \; \forall \; E \in \mathcal{B}$;

(2) $\int (af + bg) d\mu = a \int f d\mu + b \int g d\mu \; \forall f, g \in \mathcal{L}^0_+, a, b > 0$, *where* $a.\infty$
is understood as ∞;

(3) $\int f d\mu = \sup_n \int s_n d\mu$ *for any sequence* $\{s_n\} \subset \mathcal{S}_+$ *such that* $0 \leq s_n(\omega) \nearrow$
$f(\omega) \; \forall \omega \in \Omega$.

Proof: For existence, if $f \in \mathcal{L}^0_+$, we may, by Proposition 2.1.4, pick a sequence
$\{s_n\} \subset \mathcal{S}_+$ such that $0 \leq s_n(\omega) \nearrow f(\omega) \; \forall \omega \in \Omega$; define

$$\int f d\mu = \sup_n \int s_n d\mu \; .$$

By Lemma 2.2.4, this supremum (or limit) is independent of the choice of the
approximating sequence $\{s_n\}$ and depends only on f (and μ). This 'independence
of approximating sequence' and part (2) of Proposition 2.2.1 are easily seen to
imply the truth of part (2) of this proposition.

As for uniqueness, Proposition 2.2.1 implies that any 'method of integration'
which satisfies parts (1) and (2) of this proposition must agree with our method
when applied to any $s \in \mathcal{S}_+$. Part (3) and Proposition 2.1.4 then enable us to
deduce the uniqueness assertion. $\square$

Proposition 2.2.6

$$f \in \mathcal{L}^0_+ \Rightarrow \int f d\mu = \sup \left\{ \int s d\mu : s \in \mathcal{S}_+, s(\omega) \leq f(\omega) \forall \omega \in \Omega \right\} .$$

Proof: Note to start with that part (2) of Proposition 2.2.5 implies (exactly as
in the simple case) that

$$g, h \in \mathcal{L}^0_+, \; g \leq h \Rightarrow \int g d\mu \leq \int h d\mu \; . \tag{2.2.3}$$

This monotonicity of the integral immediately shows that

$$\int f d\mu \geq \sup \left\{ \int s d\mu : s \in \mathcal{S}_+, s(\omega) \leq f(\omega) \forall \omega \in \Omega \right\} .$$

Conversely, the way we have defined $\int f d\mu$ shows that it is in fact the
supremum of a countable subset of the set on the right side. $\square$

We now come to the useful and celebrated extension of Lemma 2.2.4 to
non-negative measurable (not necessarily simple) functions, the monotone
convergence theorem, which we shall refer to as simply MCT.

Theorem 2.2.7   (Monotone convergence theorem)   *If* $f, f_n \in \mathcal{L}_+^0$ *and* $f_n(\omega) \nearrow f(\omega) \forall \omega \in \Omega$, *then*

$$\int f d\mu = \sup_n \int f_n d\mu = \lim_{n \to \infty} \int f_n d\mu .$$

Proof: It follows from (2.2.3) that

$$\int f d\mu \geq \sup_n \int f_n d\mu = \lim_{n \to \infty} \int f_n d\mu . \qquad (2.2.4)$$

Conversely, suppose $s \in \mathcal{S}_+$ and $s \leq f$. Then, observe that $f_n \wedge s \nearrow s$. Since $f_n \wedge s$ is a bounded non-negative measurable function, it follows from Remark 2.1.5 that we can find $s_n \in \mathcal{S}_+$ such that

$$s_n \leq f_n \wedge s < s_n + \frac{1}{n} .$$

If we set $t_n = s_1 \vee s_2 \vee \cdots \vee s_n$, then we see that $\{t_n\} \subset \mathcal{S}_+$, $t_1 \leq t_2 \leq \cdots$ and that

$$t_n \leq f_n \wedge s < t_n + \frac{1}{n} .$$

This implies that $t_n \nearrow s$ and hence (by Proposition 2.2.5(3)) that

$$\int s d\mu = \sup_n \int t_n d\mu = \lim_n \int t_n d\mu \leq \lim_n \int f_n d\mu .$$

It follows now from Proposition 2.2.6 and the arbitrariness of s that we have

$$\int f d\mu \leq \lim_n \int f_n d\mu . \qquad (2.2.5)$$

The theorem follows from inequalities (2.2.4) and (2.2.5).   □

A useful companion to MCT is the following:

Theorem 2.2.8   (Fatou's lemma)   *If* $\{f_n : n = 1, 2, \ldots\}$ *is any sequence of non-negative measurable functions, then*

$$\int \liminf_{n \to \infty} f_n d\mu \leq \liminf_{n \to \infty} \int f_n d\mu.$$

Proof: Letting $g_n = \inf_{m \geq n} f_m$, we see that

$$0 \leq g_n(\omega) \nearrow \liminf_{n \to \infty} f_n(\omega) \ \forall \omega \in \Omega,$$

and hence, by MCT, we find that

$$\int \liminf_{n \to \infty} f_n d\mu = \lim_{n \to \infty} \int g_n d\mu.$$

Since $g_n(\omega) \le f_m(\omega) \ \forall m \ge n, \ \forall \omega \in \Omega$, we see that

$$\int g_n d\mu \le \inf_{m \ge n} \int f_m d\mu,$$

and hence that, indeed,

$$\int \liminf_{n \to \infty} f_n d\mu \le \lim_{n \to \infty} \inf_{m \ge n} \int f_m d\mu$$

$$= \liminf_{n \to \infty} \int f_n d\mu$$

as desired.   □

Notice that the above application of MCT is valid only if $\liminf_n f_n(\omega) < \infty \ \forall \omega \in \Omega$. Please see Ex. 2.2.11 for the case when the above $\liminf_n f_n$ is an extended real-valued function.

Now we proceed to make sense of the integral of 'reasonable' functions of mixed sign.

Definition 2.2.9   *Let f be a real-valued measurable function on $(\Omega, \mathcal{B}, \mu)$. If $\int f_+ d\mu < \infty$ or $\int f_- d\mu < \infty$, then we define*

$$\int f d\mu = \int f_+ d\mu - \int f_- d\mu.$$

If $\int f_\pm d\mu < \infty$; then f is said to be integrable. If $\int f_\pm d\mu = \infty$, then we say the integral of f with respect to μ does not exist.

A complex-valued measurable function is a function $f : \Omega \to \mathbb{C}$ which is $(\mathcal{B}, \mathcal{B}_\mathbb{C})$-measurable. Such an f is said to be integrable if $\int |f| d\mu < \infty$. (See Ex. 2.2.12.)

The collection of all integrable measurable functions is denoted by $\mathcal{L}^1(\Omega, \mathcal{B}, \mu)$.

Finally, we wish to see how the process of integration behaves with respect to limits. In general, nothing much can be said because of such examples as the following: let $\Omega = [0, 1]$, $\mathcal{B} = \mathcal{B}_{[0,1]}$ and μ be the Lebesgue measure on Borel subsets of $[0,1]$; let $f_n = n 1_{(0,1/n)}$: then $f_n(w) \to 0$ for all $w \in \Omega$ while $\int f_n d\mu = 1 \ \forall n$, so certainly $\lim \int f_n d\mu \ne 0$. The above example was made possible because the mass was 'allowed to escape to infinity'. In a sense, that is the only reason integration may not be well behaved with respect to limits. That 'sense' is made precise by the following extremely useful result.

Theorem 2.2.10 (**Lebesgue's dominated convergence theorem**) *Let* $\{f_n\}_{n=1}^{\infty}$ *be a sequence of integrable functions, which is uniformly dominated by an integrable function, i.e., suppose there exists* $g \in \mathcal{L}^1(\Omega, \mathcal{B}, \mu)$ *such that* $|f_n(\omega)| \le g(\omega)$ $\forall \omega \in \Omega, \forall n.$ *If* $f_n(\omega) \to f(\omega)$ $\forall \omega \in \Omega,$ *then*

$$f \in \mathcal{L}^1(\Omega, \mathcal{B}, \mu) \text{ and } \lim_{n \to \infty} \int f_n d\mu = \int f d\mu.$$

Proof: Clearly, the hypotheses imply that $|f(\omega)| \le g(\omega) \ \forall \omega \in \Omega$, so $\int |f| d\mu \le \int g d\mu < \infty$, and so f is integrable.

To prove that $\int f d\mu = \lim_{n \to \infty} \int f_n d\mu$, we may, by considering real and imaginary parts separately, assume that the f_n are real-valued. Note then that $\{g \pm f_n\}_{n=1}^{\infty}$ is sequence of non-negative measurable functions such that $\lim_{n \to \infty}(g(\omega) \pm f_n(\omega)) = g(\omega) \pm f(\omega)$. Deduce from Fatou's lemma, cf. Theorem 2.2.8, that

$$\int (g \pm f) d\mu \le \liminf \left(\int g d\mu \pm \int f_n d\mu \right).$$

It follows that

$$\int f d\mu \le \liminf \int f_n d\mu$$

and (since $\liminf(-a_n) = -\limsup a_n$) that

$$-\int f d\mu \le -\limsup \int f_n d\mu;$$

hence,

$$\limsup \int f_n d\mu \le \int f d\mu \le \liminf \int f_n d\mu.$$

Since $\limsup a_n \ge \liminf a_n$ for any sequence $\{a_n\}$ of extended real numbers, the proof is complete.   □

EXERCISES

Ex. 2.2.11   Prove Theorem 2.2.7 in the case when f_n and f are non-negative extended real-valued measurable functions. (*Hint:* Let $E_n = \{f_n = \infty\}, E = \{f = \infty\}$. Case 1: If $\mu(E_n) > 0$ for some n, then both sides of the desired identity are clearly infinite. Case 2: If $\mu(E_n) = 0 \ \forall n, \mu(\Omega \setminus E) = 0$, then apply Theorem 2.2.7 to $f_n 1_E \wedge N$ and let $N \to \infty$. In all other cases, show that the desired identity is a consequence of Theorem 2.2.7.)

Ex. 2.2.12

(1) If f is a real-valued measurable function, show that the two definitions of integrability given in Definition 2.2.9 are equivalent, i.e., $\int |f| d\mu < \infty \Leftrightarrow \int |f_+| d\mu < \infty$ and $\int |f_-| d\mu < \infty$.

(2) If f is a complex-valued measurable function, show that $f = g + ih$, where g and h are the (real-valued) measurable functions defined by $g(w) = \text{Re } f(w)$ and $h(w) = \text{Im } f(w)$. Show that f is integrable (i.e. $\int |f| d\mu < \infty$) if and only if g and h are integrable; for such an f, define $\int f d\mu = \int g d\mu + i \int h d\mu$.

(3) Show that $\mathcal{L}^1(\Omega, \mathcal{B}, \mu)$ is a complex vector space and that $\int (\cdot) d\mu : \mathcal{L}^1(\Omega, \mathcal{B}, \mu) \to \mathbb{C}$ is a complex linear functional.

Ex. 2.2.13 Let $\{f, f_n : n \geq 1\}$ be $\bar{\mathbb{R}}$-valued measurable functions on $(\Omega, \mathcal{A}, \mu)$.

(a) Assume that $f_n \uparrow f$ and that there exists an $(\mathcal{A}, \mathcal{B}_{\bar{\mathbb{R}}})$-measurable function such that $\int h_- d\mu < \infty$ and $f_n \geq h \forall n$. Then show that $\int f_n d\mu \uparrow \int f d\mu$.

(b) Show by counter-example that the above hypothesis cannot be dropped.

Ex. 2.2.14 Suppose $f : \mathbb{R}^2 \to \mathbb{R}$ is a measurable function satisfying the following conditions:

1. $f(x, \cdot)$ is integrable with respect to Lebesgue measure on $\mathbb{R}$, for each $x \in \mathbb{R}$;

2. $\frac{\partial}{\partial x} f(x, y)$ exists for all $(x, y) \in \mathbb{R}^2$; and

3. $\left| \frac{\partial}{\partial x} f(x, y) \right| \leq h(y)$ for some integrable function h on $\mathbb{R}$.

Then show that the function $\mathbb{R} \ni x \mapsto \int_{\mathbb{R}} f(x, y) dy$ is differentiable and that

$$\frac{\partial}{\partial x} \left(\int_{\mathbb{R}} f(x, y) dy \right) = \int_{\mathbb{R}} \frac{\partial}{\partial x} f(x, y) dy.$$

Ex. 2.2.15 Show that

$$\int_1^\infty e^{-s} \log s \, ds = \lim_{n \to \infty} \int_1^n \left(1 - \frac{s}{n} \right)^n \log s \, ds$$

and that

$$\int_0^1 e^{-s} \log s \, ds = \lim_{n \to \infty} \int_{\frac{1}{n}}^1 \left(1 - \frac{s}{n} \right)^n \log(s) ds.$$

2.3   a.e. Considerations

In this section, we discuss the important notion of 'things happening almost everywhere'.

Throughout this section, the symbol μ will denote a measure defined on a σ-algebra $\mathcal{B}$ of subsets of a set Ω. We shall write $\bar{\mathcal{B}} = \mathcal{M}(\mu)$ for the σ-algebra of μ-measurable sets, and we shall write $\mu(E)$ for $\mu^*(E)$ if $E \in \bar{\mathcal{B}}$; then $(\Omega, \bar{\mathcal{B}}, \mu)$ is the completion of $(\Omega, \mathcal{B}, \mu)$, i.e., $\bar{\mathcal{B}}$ contains all μ-null sets ($N \subseteq E$, $E \in \mathcal{B}$, $\mu(E) = 0 \Rightarrow N \in \bar{\mathcal{B}}$) and every member of $\bar{\mathcal{B}}$ differs by a μ-null set from a member of $\mathcal{B}$, i.e., $\mathcal{B} \in \bar{\mathcal{B}} \Rightarrow \exists K, U \in \mathcal{B} \ni K \subset B \subset U$ and $\mu(U - K) = 0$ so that $\mu(B \setminus K) = \mu(U \setminus B) = \mu(U \setminus K) = 0$. (For all this, see Ex. 1.5.15.)

Ex. 2.3.1   Let $\Omega, \mathcal{B}, \mu, \bar{\mathcal{B}}$ be as above. Suppose a function $f : \Omega \to \mathbb{R}$ is $(\bar{\mathcal{B}}, \mathcal{B}_{\mathbb{R}})$-measurable. Show that there exists a $(\mathcal{B}, \mathcal{B}_{\mathbb{R}})$-measurable function g such that $\mu(\{w : f(w) \neq g(w)\}) = 0$. (*Hint :* first consider the case when f is a simple function; next use simple approximation.)

Generally, we say that something happens **almost everywhere** if that happens outside a set of measure zero. More precisely, the content of the last exercise is that any $\bar{\mathcal{B}}$ - measurable function agrees almost everywhere with a $\mathcal{B}$-measurable function. We shall use the abbreviation a.e. to denote 'almost everywhere'. Thus, for instance, we shall say of a sequence $\{f_n\}_{n=1}^{\infty}$ of measurable functions that $f_n \to 0$ a.e. if $\mu(\{w : f_n(w) \not\to 0\}) = 0$. In view of the foregoing exercise, we may, if we wish, assume that $\mathcal{B}$ is μ-complete, i.e., that $\mathcal{B} = \bar{\mathcal{B}}$, or equivalently, that $\mathcal{B}$ contains every set of μ^*-measure 0.

Proposition 2.3.2   *If $f \in \mathcal{L}^1(\Omega, \mathcal{B}, \mu)$, then*

$$\left| \int f d\mu \right| \leq \int |f| d\mu.$$

Proof: Choose a complex number r so that $|r| = 1$ and

$$\left| \int f d\mu \right| = r \int f d\mu = \int r f d\mu.$$

Let $g = \operatorname{Re} r f$ and $h = \operatorname{Im} r f$. Then, by definition

$$\left| \int f d\mu \right| = \int g d\mu + i \int h d\mu.$$

Since the left side is real, it must be that $\int h d\mu = 0$ and that $|\int f d\mu| = \int g d\mu$. Notice now that g is a real-valued measurable function and that

$$|g| = |\operatorname{Re}(rf)| \leq |rf| = |f|,$$

and hence $|f| \pm g \geq 0$ so that also $\int |f| d\mu \pm \int g d\mu \geq 0$, or

$$\left| \int f d\mu \right| = \int g d\mu \leq \int |f| d\mu.$$

$\square$

We now re-state earlier results in the form involving a.e. considerations, in which they are used most often.

Proposition 2.3.3   *Let* $(\Omega, \mathcal{B}, \mu)$ *be a* σ*-finite measure space, and let* $f, g, h, f_1, f_2, \ldots$ *etc. denote complex-valued measurable functions on* $(\Omega, \mathcal{B})$.

(1)   *If either (a)* $f, g \in \mathcal{L}^1(\Omega, \mathcal{B}, \mu)$, *and* $\alpha, \beta \in \mathbb{C}$, *or (b)* $f, g \geq 0$, *and* $\alpha, \beta \in [0, \infty)$, *then*

$$\int (\alpha f + \beta g) d\mu = \alpha \int f d\mu + \beta \int g d\mu;$$

(2) $f \geq 0$ *a.e.* $\Rightarrow \int f d\mu \geq 0$;
(3)   *If* $f_n \geq 0$ *a.e.* $\forall n = 1, 2, \ldots$, *then*

$$\int (\liminf f_n) d\mu \leq \liminf \int f_n d\mu;$$

(4)   *If* $f_n \to f$ *a.e., and if* $|f_n| \leq g$ *a.e. for* $n = 1, 2, \ldots$, *where* $g \in \mathcal{L}^1(\Omega, \mathcal{B}, \mu)$, *then* $f, f_n \in \mathcal{L}^1(\Omega, \mathcal{B}, \mu)$ *and*

$$\lim_{n \to \infty} \int f_n d\mu = \int f d\mu;$$

(5)   *If* $0 \leq f_1(w) \leq f_2(w) \leq \ldots$ *a.e. and* $\lim_{n \to \infty} f_n(w) = f(w)$ *for* μ*-almost all* w, *i.e.,* $f_n \to f$ *a.e., then*

$$\lim_{n \to \infty} \int f_n d\mu = \int f d\mu.$$

We will not insult the reader by writing down a proof of these facts. The only things the reader should keep using in the proof are the facts that (i) if $\mu(E_n) = 0$ for $n = 1, 2, \ldots$ then $\mu(\cup_n E_n) = 0$ and (ii) if $f = 0$ a.e., then $f \in \mathcal{L}^1(\Omega, \mathcal{B}, \mu)$ and $\int f d\mu = 0$. There is a partial converse to (ii) which is useful.

Proposition 2.3.4   *Let* f *be a non-negative measurable function. Then, the following conditions on* f *are equivalent:*

(1) $\int f d\mu = 0$
(2) $f = 0$ *a.e.*

Proof: Suppose $f = 0$ a.e. If g is a non-negative simple measurable function such that $0 \leq g \leq f$, and if $g = \sum_{i=1}^{n} a_i 1_{E_i}$, where $a_1, \ldots, a_n$ are the distinct non-zero values in the range of g, then $\mu(E_i) = 0$ for all i (since $E_i \subseteq \{f \neq 0\}$) and hence $\int g d\mu = \sum_{i=1}^{n} a_i 0 = 0$. Hence $\int f d\mu = 0$.

Conversely, suppose $f \geq 0$ and $\int f d\mu = 0$. Let $E_n = \{f > 1/n\}$. Then $f \geq \frac{1}{n}1_{E_n}$ and hence

$$\frac{1}{n}\mu(E_n) = \int \frac{1}{n}1_{E_n}d\mu \leq \int f d\mu = 0$$

so that $\mu(E_n) = 0$ for all n. Then

$$\mu\left(\bigcup_{n=1}^{\infty}E_n\right) = 0, \text{ but } \bigcup_{n=1}^{\infty}E_n = \{f \neq 0\};$$

in other words $f = 0$ outside a set of zero measure.   □

We shall sometimes find it convenient to write $\int_E f d\mu$ for $\int 1_E f d\mu$, whenever f is either non-negative or integrable.

EXERCISES

Ex. 2.3.5 Let f be a non-negative measurable function defined on the measure space $(\Omega, \mathcal{B}, \mu)$. Define

$$\mu_f : \mathcal{B} \to [0, \infty] \text{ by } \mu_f(E) = \int_E f d\mu.$$

Show that:

(1) μ_f is a measure defined on $\mathcal{B}$.
(2) μ_f is σ-finite if and only f is finite almost everywhere.
(3) $E \in \mathcal{B}, \mu(E) = 0 \Rightarrow \mu_f(E) = 0$.
    The measure μ_f is sometimes called the 'indefinite integral of f'.

Ex. 2.3.6 Prove that if $f \in \mathcal{L}^1(\Omega, \mathcal{B}, \mu)$, the following are equivalent:

(1) $\int_E f d\mu = 0$ for all $E \in \mathcal{B}$.
(2) $f = 0$ μ a.e.

3

Random Variables

3.1 Distribution and Expectation

Let $(\Omega, \mathcal{B}, P)$ be a probability space. Thus, P is a measure on $\mathcal{B}$ and $P(\Omega) = 1$.

Definition 3.1.1 *A random variable X on a probability space $(\Omega, \mathcal{B}, P)$ is a real-valued $\mathcal{B}$-measurable function on Ω.*

Definition 3.1.2 *If $(\Omega, \mathcal{B}, P)$ is a probability space and X is a random variable on Ω, the **distribution** of the random variable X is, by definition, the probability measure P_X defined on $(\mathbb{R}, \mathcal{B}_{\mathbb{R}})$ by:*

$$P_X(E) = P(X^{-1}(E)) = P(\{w \in \Omega : X(w) \in E\})$$

$$= P(X \in E),$$

for any $E \in \mathcal{B}_{\mathbb{R}}$.

The function given by $F_X(x) = P(X \leq x)$, $x \in \mathbb{R}$ determines the measure P_X and is called the distribution function of X. (See Ex. 1.6.6 for this and properties of F_X.)

Definition 3.1.3 *A random variable X on $(\Omega, \mathcal{B}, P)$ is said to be*

(1) **discrete** *if there exists a countable set $C = \{x_i : i \in \mathbb{N}\}$ in $\mathbb{R}$ such that $P_X(C) = 1$, or equivalently*

$$P_X(E) = \sum_{x_i \in E \cap C} P(X = x_i) \quad \forall E \in \mathcal{B}_R,$$

(2) **continuous** *if the distribution function $F_X(x) = P(X \leq x)$ is a continuous function,*

(3) **absolutely continuous** *if*

$$P_X(E) = \int_E f(x)dx \quad \forall E \in \mathcal{B}_\mathbb{R}$$

for a non-negative Borel-measurable function $f: \mathbb{R} \rightarrow \mathbb{R}$, *which is referred to as the density of* X.

From the above definition, it is easy to see that X is discrete if $F_X(\cdot)$ is a step function.

Definition 3.1.4   *A function* $f : \mathbb{R} \rightarrow \mathbb{R}$ *is said to be* **absolutely continuous** *if* $\forall \epsilon > 0$ *there exists* $\delta > 0$ *such that*

$$\sum_{i=1}^n |f(x_i) - f(y_i)| < \epsilon,$$

for all finite collections of pairwise disjoint intervals (x_i, y_i) *with* $\sum_{i=1}^n (y_i - x_i) < \delta$.

We shall see later (Section 7.5) that $F_X(\cdot)$ is absolutely continuous if and only if X is an absolutely continuous random variable, and, in this case, $\frac{d}{dx} F_X(x) = f(x)$ a.e. There are examples of distribution functions which are continuous but not absolutely continuous (see Ex. 7.5.14).

Definition 3.1.5   *Let* X *be a random variable on a probability space* $(\Omega, \mathcal{B}, P)$. *We write* $E(X)$ *for* $\int X dP$ *when this makes sense, and call* $E(X)$ '*the* **expectation of** X'.

The following special case of dominated convergence is usually referred to as Bounded Convergence Theorem whose obvious proof we omit.

Theorem 3.1.6   *Let* $X, X_n, n \in \mathbb{N}$ *be a sequence of random variables on a probability space* $(\Omega, \mathcal{B}, P)$. *Suppose* $X_n \rightarrow X$ *a.e. and there exists* $K > 0$ *such that* $|X_n| \leq K < \infty$ *a.e. Then* $EX_n \rightarrow EX$.

We conclude this section with a standard inequality.

Proposition 3.1.7   **(Tchebychev's inequality)** *Let* X, Y *be two random variables on a probability space* $(\Omega, \mathcal{B}, P)$. *Then* $\forall \lambda > 0$,

$$P(|X| > \lambda) \leq \frac{E(|X|)}{\lambda}. \tag{3.1.1}$$

Proof: As $|X|$ is a non-negative random variable, $E(|X|)$ makes sense. Note that

$$E(|X|) = \int_{|X|>\lambda} |X| dP + \int_{|X|\leq\lambda} |X| dP$$

$$\geq \int_{|X|>\lambda} |X| dP$$

$$\geq \int_{|X|>\lambda} \lambda dP$$

$$= \lambda P(|X| > \lambda).$$ □

EXERCISES

Ex. 3.1.8   Let $(\Omega, \mathcal{B}, P)$ be a probability space. Suppose

(1) X is discrete, with range $\{x_i : i \in \mathbb{N}\}$ and $g : \mathbb{R} \to \mathbb{R}$. Then $E(g(X)) = \sum_{i=1}^{\infty} g(x_i)P(X = x_i)$, provided $\sum_{i=1}^{\infty} |g(x_i)|P(X = x_i) < \infty$.
(2) X is absolutely continuous with density f and $g : \mathbb{R} \to \mathbb{R}$. Then $E(g(X)) = \int g(x)f(x)dx$ provided $\int |g(x)|f(x)dx < \infty$.

Ex. 3.1.9   The **moment generating function** of a random variable X is defined to be the function $M_X(t) = E(e^{tX}) = \sum_{n=0}^{\infty} \frac{E(X^n)}{n!}t^n$. Let $I = \{t \in \mathbb{R} : M_X(t) < \infty\}$. Show that

(1) I is a possibly degenerate interval and $0 \in I$.
(2) $M_X(\cdot)$ is a continuous convex function on I.
(3)   if 0 is an interior point of I, then $E(X^k) < \infty$ for all $k \in \mathbb{N}$ (i.e. X has finite moments of all orders).

Ex. 3.1.10   Let X be a random variable on the probability space $(\Omega, \mathcal{B}, P)$, with distribution P_X. Consider the random variable $\widetilde{X}$ on the probability space $(\mathbb{R}, \mathcal{B}_\mathbb{R}, P_X)$ defined by $\widetilde{X}(x) = x$. Then $P_{\widetilde{X}} = P_X$.

Ex. 3.1.11   Let $F : \mathbb{R} \to [0, 1]$ be a distribution function of a probability measure P (i.e., $F(x) = P((-\infty, x])$). Then, show that there is a random variable $X : ((0, 1], \mathcal{B}, \lambda) \to \mathbb{R}$, (where $\mathcal{B}$ is the Borel σ-algebra and λ is Lebesgue measure), such that $P_X = P$.

Ex. 3.1.12   Let X: $\Omega \to \mathbb{N}$ be a random variable on a probability space $(\Omega, \mathcal{B}, P)$. Show that

$$E(X) = \sum_{n=1}^{\infty} P(X \geq n).$$

3.2   Independent Events and Tail σ-Algebra

Let $(\Omega, \mathcal{B}, P)$ be a probability space. Intuitively speaking, we say that two events $A, B \in \mathcal{B}$ are independent if the occurrence of one has no effect on the occurrence of the other.

Definition 3.2.1

(1)   *A finite collection $A_i \in \mathcal{B}, 1 \leq i \leq n$ of events is said to be independent if*
$$P(\cap_{i=1}^n A_i) = \prod_{i=1}^n P(A_i)$$

(2)   *A finite collection of σ-algebras, $\mathcal{B}_i, 1 \leq i \leq n$, contained in $\mathcal{B}$ are said to be independent if $P(\cap_{i=1}^n B_i) = \prod_{i=1}^n P(B_i), \forall B_i \in \mathcal{B}_i, 1 \leq i \leq n$.*

(3)   *A finite collection $X_1, \dots, X_n$ of random variables defined on a probability space $(\Omega, \mathcal{B}, P)$ is said to be independent if*

$$P(\cap_{i=1}^n (X_i \in A_i)) = \prod_{i=1}^n P(X_i \in A_i)$$

for arbitrary Borel sets $A_1, \dots, A_n$ in $\mathbb{R}$, or equivalently if the σ-algebras $\sigma(X_i), 1 \leq i \leq n$ are independent.

(4)   *An arbitrary collection of events (resp. σ-algebras, random variables) is said to be independent if every finite sub-collection is independent.*

We now prove the Borel–Cantelli Lemma a powerful tool in probability. (We shall see an application of it in Lemma 5.2.4.) This lemma is concerned with events of the form 'A_n occurs infinitely often (i.o.)', by which is meant $\{\omega : \omega \in A_n$ for infinitely many $n\}$.

Lemma 3.2.2   (Borel–Cantelli lemma) *Let $(\Omega, \mathcal{B}, P)$ be a probability space. Let $\{A_n\}$ be a sequence of events in $\mathcal{B}$. Then, the following hold:*

(1) $\sum_{n=1}^\infty P(A_n) < \infty$ *implies $P(A_n$ occurs i.o.$) = 0$.*

(2)   *If, in addition, the A_n are independent, then $\sum_{n=1}^\infty P(A_n) = \infty$ implies $P(A_n$ occurs i.o.$) = 1$.*

Proof:

(1)   Let $B_n = \cup_{k=n}^\infty A_k$. Note that $B_n \downarrow \cap_{n=1}^\infty B_n$. Hence,

$$P(A_n \text{ occurs i.o.}) = \lim_{n \to \infty} P(B_n)$$

$$= \lim_{n \to \infty} P(\cup_{k=n}^\infty A_k)$$

$$\leq \lim_{n \to \infty} \sum_{k=n}^{\infty} P(A_k)$$

$$= 0,$$

where the last equality follows from the hypothesis.

(2)   First, we note that it is enough to show $P(\cup_{n=1}^{\infty} \cap_{k=n}^{\infty} A_k^c) = 0$. Now, for $m \geq n$, using the independence hypothesis,

$$P(\cap_{k=n}^{m} A_k^c) = \prod_{k=n}^{m} P(A_k^c)$$

$$= \prod_{k=n}^{m} (1 - P(A_k))$$

$$\leq \prod_{k=n}^{m} e^{-P(A_k)}$$

$$= e^{-\sum_{k=n}^{m} P(A_k)}, \tag{3.2.2}$$

where we have used Ex. 3.2.6 in the first step and the inequality $1 - x \leq e^{-x}$ for $x \geq 0$ in the third step. (Verify the stated inequality.) Observe that

$$P(\cup_{n=1}^{\infty} \cap_{k=n}^{\infty} A_k^c) = \lim_{n \to \infty} \lim_{m \to \infty} P(\cap_{k=n}^{m} A_k^c)$$

Hence, from (3.2.2) and the hypothesis, the claim follows.   □

The following example shows that the independence assumption in the second part of the Borel–Cantelli lemma cannot be dropped.

Example 3.2.3   Let $\Omega = [0, 1]$ and P be the Lebesgue measure. Define $A_n = [0, \frac{1}{n}]$. It is easy to see that A_n are not independent and $\sum_{n=1}^{\infty} P(A_n) = \infty$ but $P(A_n$ occurs i.o. $) = P(0) = 0$.

In the Borel–Cantelli lemma, we showed that if A_n are independent events, then $P(\{A_n$ occur i.o. $\})$ is 0 or 1. It turns out that such events belong to a special σ-algebra whose events must necessarily have probability 0 or 1.

Definition 3.2.4   *Let $X_1, X_2, X_3, \ldots$ be a sequence of random variables on a probability space $(\Omega, \mathcal{B}, P)$. Then the σ-algebra*

$$\mathcal{T} = \cap_{n=1}^{\infty} \sigma(X_n, X_{n+1}, \ldots)$$

is called the **tail** σ-**algebra** *associated with the sequence* $\{X_n\}$. *An event A is a* **tail event** *if it is in the tail σ-algebra.*

Lemma 3.2.5 **(Kolmogorov's zero–one law)** *Let A be a tail event associated with the sequence of independent random variables* $\{X_n\}$. *Then* $P(A) = 0$ *or* $P(A) = 1$.

Proof: We shall show that the event A is independent of itself. This will force $P(A) = P(A \cap A) = P(A)^2$, which will imply the result.

Now $A \in \mathcal{T}$ implies that $A \in \sigma(X_n, X_{n+1}, \ldots)$ for every $n \geq 1$. Consequently, this will imply that for any $A_i \in \sigma(X_i)$, A is independent of $A_1, A_2, \ldots, A_{n-1}$ for all $n \geq 1$ (why ?). Hence $\sigma(A)$ and $\sigma(X_1, X_2, \ldots)$ are independent; however, $A \in \mathcal{T} \subset \sigma(X_1, X_2, \ldots)$. Therefore A is independent of itself. $\square$

EXERCISES

Ex. 3.2.6   Show that the following are equivalent: (i) A family A_i of events is independent; (ii) The family $\sigma(1_{A_i})$ of σ-algebras is independent.

Ex. 3.2.7   Let X, Y be random variables on a probability space $(\Omega, \mathcal{B}, P)$. Show that X and Y are independent if and only if $\sigma(X)$ and $\sigma(Y)$ are independent.

3.3   Some Distributions

3.3.1   Discrete probability distributions

In this section, we shall discuss some discrete random variables. There are two important parameters associated with such random variables X. The first is the expectation (or **mean**), $m = E(X) = \int X \, dP = \sum_{i=1}^{\infty} x_i P(X = x_i)$, provided the sum converges absolutely. If we think of P_X as a mass distribution on the real line, then the number m denotes the centre of gravity of this mass distribution. The second is the **variance** or the mean square deviation, $\sigma^2 = E(X - m)^2 = E(X^2) - m^2$. The number σ is referred to as the standard deviation X. The variance signifies the spread of the distribution P_X about the mean m.

We will now discuss distributions of some frequently encountered discrete random variables. We will not mention the precise probability spaces on which the random variables are defined but these can be easily constructed by the reader.

- **Bernoulli distribution** (p): Consider the following experiment. Toss a coin and note the outcome. Assume that the chance of a head appearing in the toss is p. Let the random variable X assume the value 0 if a tail appears and 1 if a head appears. It is easy to see that the distribution of X is given by

$$P_X(\{x\}) = P(X = x) = p^x(1 - p)^{1-x}, \ x = 0, 1,$$

with mean $m = p$ and variance $\sigma^2 = p(1 - p)$. The random variable X is said to be distributed as Bernoulli (p) and is denoted by $X \overset{d}{=}$ Bernoulli (p).

- **Binomial distribution** (n, p): Consider the following experiment. Toss a coin n times so that each toss is independent of the others and the chance of getting a head is p. Let X be the total number of heads in n tosses. It is easy to see that the range $(X) = \{0, 1, 2, \ldots, n\}$ and that the distribution of X is given by

$$P_X(\{k\}) = P(X = k) = \binom{n}{k} p^k (1 - p)^{n-k}, k = 0, 1, \ldots, n,$$

with mean $m = np$ and variance $\sigma^2 = np(1 - p)$. The random variable X is said to be distributed as Binomial (n, p) and this is denoted by $X \overset{d}{=}$ Binomial (n, p).

- **Geometric distribution** (p) Consider tossing a coin till the first head appears. Assume that each toss is independent of the others and the chance of getting a head is p. Let X be the trial at which the first head appears. It is easy to see that the range$(X) = \mathbb{N}$ and that the distribution of X is given by

$$P(X = k) = p(1 - p)^{k-1}, k \in \mathbb{N},$$

with mean $m = \frac{1}{p}$ and variance $\sigma^2 = \frac{(1-p)}{p^2}$. The random variable X is said to be distributed as Geometric (p) and is denoted by $X \overset{d}{=}$ Geometric (p).

- **Uniform distribution** $\{1, \ldots, n\}$: Consider an experiment in which we choose a number from $\{1, \ldots, n\}$ at random (i.e. every number has an equally likely chance of being chosen). Let X be the number chosen. It is easy to see that range $(X) = \{1, \ldots, n\}$ and that the distribution of X is given by

$$P(X = k) = \frac{1}{n}, 1 \le k \le n,$$

with mean $m = \frac{n+1}{2}$ and variance $\sigma^2 = \frac{n^2-1}{12}$. The random variable X is said to be distributed as Uniform distribution $\{1, \ldots, n\}$ and this is denoted by $X \overset{d}{=}$ Uniform $(\{1, 2, \ldots, n\})$.

- **Poisson distribution** (λ) Let $\lambda > 0$. Consider a random variable X whose range is $\mathbb{N} \cup \{0\}$ and whose distribution is given by

$$P_X(\{k\}) = e^{-\lambda}\frac{\lambda^k}{k!}, k = 0, 1, 2, 3, \dots .$$

This distribution arises as a limiting case of the Binomial (n, p) when $n \to \infty$, $p \to 0$, with $\lambda = np$ being constant. This is illustrated in Ex. 3.3.5. It is easy to check that its mean $m = \lambda$ and variance $\sigma^2 = \lambda$. The random variable X is said to be distributed as Poisson (λ) and is denoted by $X \overset{d}{=}$ Poisson (λ).

EXERCISES

Ex. 3.3.1   Verify that the mean and the variance of Bernoulli (p), Binomial (n, p), Geometric (p), Uniform $\{1, 2, \dots, n\}$ and Poisson (λ) given above are correct.

Ex. 3.3.2   Suppose $X_1, X_2, \dots, X_n$ are independent Bernoulli (p) random variables. Then find the distribution of $S_n = \sum_{k-1}^{n} X_i$.

Ex. 3.3.3   Consider independent Bernoulli (p) trials. Let Y be a random variable that denotes the trial at which the rth head appears.

(1)   Find the distribution of Y.
(2) Y is said to be distributed as Negative Binomial (r, p). Calculate its mean and variance.

Ex. 3.3.4   Suppose you are given a bag of N balls, of which g are good and b are bad. You choose $n \leq N$ balls from the bag at random.

(1)   Let Y be the number of good balls chosen amongst the n balls. Find the distribution of Y.
(2) Y is said to be distributed as Hypergeometric (N, n, g). Calculate its mean and variance.

Ex. 3.3.5   Let $\lambda > 0$, let $(\Omega, \mathcal{B}, P)$ be a probability space such that the random variables, $X_n \overset{d}{=}$ Binomial $(n, \frac{\lambda}{n})$. Show that

$$\lim_{n \to \infty} P(X_n = k) = e^{-\lambda}\frac{\lambda^k}{k!}, \forall k \in 0, 1, \dots$$

Ex. 3.3.6   Consider an experiment in which there are k possible outcomes, and outcome $i, 1 \leq i \leq k$ occurs with probability p_i. Repeat this experiment n

times, with each trial independent of the others. Let Y_i be the number of times that outcome i appears in n trials and let $Y = (Y_1, Y_2, \ldots, Y_k)$. Show that

$$P(Y_1 = y_1, \ldots, Y_k = y_k) = \frac{n!}{y_1!, \ldots, y_k!} p_1^{y_1}, \ldots, p_k^{y_k},$$

where $\sum_{i=1}^{k} y_k = n$ and $\sum_{i=1}^{k} p_i = 1$. Y is said to be distributed as Multinomial $(k, n, p_1 \ldots p_k)$.

Ex. 3.3.7   Suppose two teams play a series of games, each producing a winner and a loser until one team has won two games more than the other. Let G be the total number of games played. Assuming that each team has chance 0.5 to win each game independent of the results of the previous games, find $E(G)$.

3.3.2   Absolutely continuous probability distributions

In this section, we shall discuss some frequently encountered random variables which are absolutely continuous. Let X be an absolutely continuous random variable with density f. As before, the two important parameters associated with X are the mean $m = E(X) = \int x f(x) dx$ and the variance $\sigma^2 = \int (x - m)^2 f(x) dx$ provided these integrals exist.

- **Uniform distribution on** (a, b): This distribution is intended to simulate the following experiment: choose a point at random in (a, b), $-\infty < a < b < \infty$. The main assumption is that chance that a chosen point is in a given interval is proportional to its length. Let X be a random variable denoting the chosen point. Then X is said to be distributed Uniform (a, b) if its density is given by:

$$f(x) = \begin{cases} \frac{1}{b-a} & a < x < b \\ 0 & \text{otherwise.} \end{cases}$$

  Note that for $a < c < d < b$, $P(X \in (c, d)) = \int_c^d f(x) dx = \frac{d-c}{b-a}$ (which agrees with our main assumption) and that the mean $m = \frac{b+a}{2}$ and variance $\sigma^2 = \frac{(b-a)^2}{12}$.

- **Exponential distribution** (λ):   A random variable T is said to have Exponential (λ) distribution if its density is given by

$$f(t) = \begin{cases} \lambda e^{-\lambda t} & t \geq 0 \\ 0 & \text{otherwise.} \end{cases}$$

  Its mean $m = \frac{1}{\lambda}$, and variance $\sigma^2 = \frac{1}{\lambda^2}$. This distribution has been found to be a good model for the lifetime of a bulb as well as the time until decay of a radioactive particle.

- **Normal distribution** (m, σ^2):   This distribution is, as the name indicates, ubiquitous in, and central to, probability. Its significance will become clear soon. A random variable X is said to be normally distributed, denoted by $N(m, \sigma^2)$, if its density is given by

$$f(x) = \frac{1}{\sqrt{2\pi}\sigma} e^{-\frac{(x-m)^2}{2\sigma^2}}, \quad -\infty < x < \infty.$$

It can be shown that $E(X) = m$ and $\text{Var}(X) = \sigma^2$.

- **Standard Normal distribution and its properties**   The random variable $Z \overset{d}{=} N(0, 1)$ is said to be standard Normal. Then the distribution function of Z, denoted by Φ, is given by $\Phi(a) = \frac{1}{\sqrt{2\pi}} \int_{-\infty}^{a} e^{\frac{-z^2}{2}} dz$ for all $a \in \mathbb{R}$.

  (1)   By symmetry of the density of the standard Normal distribution, it can be easily seen that

$$\Phi(a) = 1 - \Phi(-a) \ \forall a \in \mathbb{R}.$$

  (2)   If $X = \sigma Z + m$, where $\sigma > 0$ and $m \in \mathbb{R}$, then

$$P(X \le x) = P\left(Z \le \frac{x - m}{\sigma}\right)$$

$$= \frac{1}{\sqrt{2\pi}} \int_{-\infty}^{\frac{x-m}{\sigma}} e^{\frac{-z^2}{2}} dz$$

$$= \frac{1}{\sqrt{2\pi}\sigma} \int_{-\infty}^{x} e^{-\frac{(z-m)^2}{2\sigma^2}} dz.$$

  So $X \overset{d}{=} N(m, \sigma^2)$. Moreover it is easy to verify that

$$P(a \le X \le b) = \Phi\left(\frac{b - m}{\sigma}\right) - \Phi\left(\frac{a - m}{\sigma}\right).$$

  By a similar argument in reverse, we can conclude that $X \overset{d}{=} N(m, \sigma^2)$ if and only if $Z = \frac{X-m}{\sigma}$ has standard Normal distribution.

  (3)   In general, it is impossible to exactly evaluate the integral defining $\Phi(\cdot)$. So numerical values of the integral are available, using which one can calculate the values. The Normal tables are given in Table B.2 (see the Appendix).

EXERCISES

Ex. 3.3.8 Verify that the mean and the variance of Uniform (a, b), Exponential (λ) and Normal (m, σ^2) given above are correct.

Ex. 3.3.9 Suppose X has Uniform $[-2, 4]$ distribution. Find the density of X^2.

Ex. 3.3.10 Beginning at midnight, a computer centre is up for one hour and down for 2 hours in a regular cycle. A person who does not know the schedule dials the centre at a random time uniformly selected between 12:00 midnight and 5:00 am. What is the probability that the centre will be operating when this call comes in?

Ex. 3.3.11 Let $\{X_k : 1 \le k \le n\}$ be independent, continuous, random variables with identical distribution function F. Find the probability density function of $\max(X_1, \ldots, X_n)$ in terms of F and its derivatives.

Ex. 3.3.12 Let $\{X_k : 1 \le k \le n\}$ be independent continuous random variables identically distributed as Uniform $(0, 1)$. Let

$$X_{(1)} \le X_{(2)} \le \cdots \le X_{(k)} \le \cdots \le X_{(n)}$$

be the increasing rearrangement of the random variables $\{X_k : 1 \le k \le n\}$. That is, $X_{(1)}$ is the smallest of $\{X_k : 1 \le k \le n\}$, $X_{(2)}$ is the next smallest and so on. Show that the density of $X_{(k)}$ is given by

$$f_{(k)}(x) = \frac{1}{B(k, n - k + 1)} x^{k-1} (1 - x)^{n-k}, \quad 0 < x < 1,$$

where $B(r, s) = \int_0^1 x^{r-1}(1 - x)^{s-1} dx$ for any $r, s > 0$. $X_{(k)}$ is referred to as the kth order statistic and is said to have Beta $(k, n - k + 1)$ distribution.

Ex. 3.3.13 Let $X_1, X_2, \ldots, X_n$ be independent Exponential (λ) random variables. Then show that $S_n = \sum_{k=1}^{n} X_k$ has a density f_n given by

$$f_n(t) = t^{n-1} e^{-\lambda t}, \quad t \ge 0.$$

S_n is said to be distributed as Gamma (n, λ).

3.4 Conditional Expectation

In this section, we will discuss the concept of conditional expectation of a random variable. This idea will be used extensively later.

Theorem 3.4.1   *Let X be an integrable random variable on a probability space $(\Omega, \mathcal{B}, P)$ and $\mathcal{C}$ be a sub-σ-algebra of $\mathcal{B}$. Then there exists an essentially unique $\mathcal{C}$ measurable random variable Y such that*

$$\int_{\mathcal{C}} Y \, dP = \int_{\mathcal{C}} X \, dP \ \forall C \in \mathcal{C}.$$

This random variable Y is called the **conditional expectation of X given $\mathcal{C}$** *and denoted by $E(X \mid \mathcal{C})$.*

Proof: Uniqueness (up to a.e. equivalence) is easy to see: if there are two such random variables, say Y_1 and Y_2, then the fact that $\int_A Y_1 \, dP = \int_A Y_2 \, dP$ for all $A \in \mathcal{C}$ implies that $Y_1 = Y_2$ a.e.

Suppose X is a non-negative random variable. Define a measure on $(\Omega, \mathcal{C})$ by

$$Q(A) = \int_A X \, dP \ \forall A \in \mathcal{C}.$$

It is easy to see that Q and the measure P restricted to $\mathcal{C}$ satisfy the hypothesis of the Radon–Nikodym theorem (Theorem 7.3.2). Hence, there exists a random variable Y on $(\Omega, \mathcal{C})$ such that $Q(A) = \int_A Y \, dP \ \forall A \in \mathcal{C}$. This proves the theorem for a non-negative X. The case of a general X is achieved by using linearity. □

Remark 3.4.2   *From the uniqueness assertion in the above theorem it follows that if X is $\mathcal{C}$ measurable, then $E(X \mid \mathcal{C}) = X$.*

Conditional probabilities: If $\Omega = \coprod_{i=1}^{n} B_i$ and $\mathcal{C} = \sigma(\{B_1, \dots, B_n\})$, then it is not hard to see that a function $f : \Omega \to \mathbb{R}$ is $\mathcal{C}$-measurable if and only if it is a linear combination of $\{1_{B_i} : 1 \leq i \leq n\}$. Let us assume, without loss of generality, that $P(B_i) > 0$ for all $1 \leq i \leq n$.

It then follows that

$$E(1_A \mid \mathcal{C}) = \sum_{i=1}^{n} c_i 1_{B_i}$$

for appropriate constants, which are easily seen to be given by

$$c_i = \frac{P(A \cap B_i)}{P(B_i)}.$$

In elementary books on probability course, one might have seen the so-called 'conditional probability of A given B' defined for any two events A, B with

$P(B) > 0$, by the equation

$$P(A \mid B) = \frac{P(A \cap B)}{P(B)}.$$

(Thus the c_i of the preceding paragraph are nothing but $P(A \mid B_i)$).

Now we shall pass to conditional expectation with respect to general (not necessarily finite) σ-subalgebras C.

Proposition 3.4.3   *Let X be an integrable random variable on a probability space (Ω, B, P) and C be a sub-σ-algebra of B.*

(1)   *If $\sigma(X)$ is independent of C, then*

$$E(X \mid C) = E(X). \tag{3.4.3}$$

(2)   *Let Y be a random variable on (Ω, C) such that XY is integrable, then*

$$E(XY \mid C) = YE(X \mid C). \tag{3.4.4}$$

(3)   *If D is another sub-σ-algebra, $D \subset C \subset B$, then*

$$E(E(X \mid C) \mid D) = E(X \mid D) = E(E(X \mid D) \mid C). \tag{3.4.5}$$

Proof:

(1)   Suppose $X = 1_B$ for some $B \in B$. Let $A \in C$, then

$$\int_A X dP = P(A \cap B)$$

$$= P(A)P(B) \text{ (independence)}$$

$$= \int_A E(X) dP.$$

So we have shown the result when $X = 1_B$ for some $B \in B$. The same argument implies that the result is true for any non-negative simple function. For any non-negative random variable X independent of C, we can construct a sequence of simple functions increasing to X that are also independent of C (why?). Then the standard simple function approximation argument completes the proof.

(2)   Consider $Y = 1_A$ for some $A \in \mathcal{C}$. Then, for any $C \in \mathcal{C}$,

$$\int_C YE(X \mid C)dP = \int_{A \cap C} E(X \mid C)dP$$

$$= \int_{A \cap C} XdP \ (\text{by definition})$$

$$= \int_C X1_A dP$$

$$= \int_C XYdP.$$

As $YE(X \mid \mathcal{C})$ is $\mathcal{C}$-measurable, we can conclude the result when $Y = 1_A$ for some $A \in \mathcal{C}$. By a similar argument of linearity and simple function approximation as in (1), we deduce the result for any $\mathcal{C}$ measurable Y.

(3) $E(X \mid \mathcal{D})$ is a $\mathcal{D}$-measurable map and, hence, is $\mathcal{C}$-measurable. Therefore, the right equality holds by Remark 3.4.2. For the left equality, consider any $A \in \mathcal{D}$. Note that $A \in \mathcal{C}$ as well. So,

$$\int_A E(E(X \mid \mathcal{C}) \mid \mathcal{D})dP = \int_A E(X \mid \mathcal{C})dP$$

$$= \int_A XdP$$

$$= \int_A E(X \mid \mathcal{D})dP. \qquad \square$$

EXERCISES

Ex. 3.4.4   Suppose (X, Y) is a point chosen uniformly on the triangle $\{(x, y): x \geq 0, y \geq 0, x + y = 4\}$. Find the conditional probability $P(Y > 1 \mid X = x)$.

Ex. 3.4.5   Let X, X_n, $n \in \mathbb{N}$, be integrable random variables on a probability space, $(\Omega, \mathcal{B}, P)$ and $\mathcal{C}$ be a sub-σ algebra of $\mathcal{B}$.

(1)   Suppose X_n, $X \geq 0$ and $X_n \uparrow X$ on Ω. Then show that

$$E(X_n \mid \mathcal{C}) \uparrow E(X \mid \mathcal{C}).$$

(2)   Suppose $X_n \to X$ such that there exists an integrable Y on $\mathcal{C}$ such that $\mid E(X_n \mid \mathcal{C}) \mid \leq Y \ \forall n$; then

$$E(X_n \mid \mathcal{C}) \to E(X \mid \mathcal{C}).$$

4

Probability Measures on Product Spaces

4.1 Product Measures

Just as the Lebesgue measure on $\mathbb{R}^2$ is related to the Lebesgue measure on $\mathbb{R}^1$ (through the fact that the area of a rectangle is the product of the lengths of the sides), we can, given σ-finite measure spaces $(\Omega_i, \mathcal{B}_i, \mu_i)$, $i = 1, 2$, construct the 'product measure space' $(\Omega_1 \times \Omega_2, \mathcal{B}_1 \otimes \mathcal{B}_2, \mu_1 \times \mu_2)$ thus:

Let $\Omega = \Omega_1 \times \Omega_2$; call a set of the form $A_1 \times A_2 \subseteq \Omega$, $A_i \subseteq \Omega_i$ a rectangle, and call such a rectangle measurable if $A_i \in \mathcal{B}_i$, $i = 1, 2$. Define $\mathcal{B}_1 \otimes \mathcal{B}_2 = \sigma(\mathcal{R})$, where $\mathcal{R}$ is the collection of **measurable rectangles**.

Ex. 4.1.1 With the preceding notation, show that $\mathcal{A}(\mathcal{R})$, the algebra generated by measurable rectangles, is precisely the collection of sets of the form $\bigsqcup_{i=1}^{n} R_i$, $R_i \in \mathcal{R}$.

Our first step is:

Proposition 4.1.2 *If* $(\Omega_1, \mathcal{B}_1, \mu_1)$ *and* $(\Omega_2, \mathcal{B}_2, \mu_2)$ *are* σ-*finite measure spaces, then there exists a unique* σ-*finite measure* μ *defined on the* σ-*algebra* $\mathcal{B}_1 \otimes \mathcal{B}_2$ *generated by* $\mathcal{R} = \{A_1 \times A_2 : A_i \in \mathcal{B}_i\}$ *such that* $\mu(A_1 \times A_2) = \mu_1(A_1)\mu_2(A_2)$ *whenever* $A_1 \in \mathcal{B}_1$ *and* $A_2 \in \mathcal{B}_2$.
The measure μ *is denoted by* $\mu_1 \times \mu_2$ *and referred to as the* **product measure.**

One way to attempt to prove the theorem would be to define μ on rectangles as the product of the measures of the sides, and to attempt to invoke the Caratheodory extension theorem, for which, however, we have to verify countable additivity of μ on $\mathcal{A}(\mathcal{R})$ (which, in turn, would lead us to seek some topological assumptions about the Ω_i).

The method we shall adopt will neatly circumvent the need for such topological considerations, and it works in complete generality.

The key notion in the proposed method is that of slices or sections (of a subset of, or a function on a product space). If $E \subseteq \Omega_1 \times \Omega_2$, the x-**slice** (resp., y-**slice**) of E is defined by $E^x = \{y \in \Omega_2 : (x, y) \in E\}$ (resp., $E_y = \{x \in \Omega_1 : (x, y) \in E\}$. (Since subsets correspond to $\{0, 1\}$-valued functions, the last definition may be viewed as a special case of the next.) If $f : \Omega_1 \times \Omega_2 \to \mathbb{C}$ is a function, define $f^x : \Omega_2 \to \mathbb{C}$ (resp., $f_y : \Omega_1 \to \mathbb{C}$) by $f^x(y) = f(x, y)$ (resp., $f_y(x) = f(x, y)$).

The crucial observation is that if $E = A_1 \times A_2$, $A_i \in \mathcal{B}_i$, then the following assertions are valid:

(1) if $x \in \Omega_1$, then $E^x \in \mathcal{B}_2$, and the function $x \to \mu_2(E^x)$ is $(\mathcal{B}_1, \mathcal{B}_\mathbb{R})$-measurable;

(2) if $y \in \Omega_2$, then $E_y \in \mathcal{B}_1$, and the function $y \to \mu_1(E_y)$ is $(\mathcal{B}_2, \mathcal{B}_\mathbb{R})$-measurable; and

(3) $\int \mu_2(E^x) d\mu_1(x) = \int \mu_1(E_y) d\mu_2(y)$.

Proof of Proposition 4.1.2:   It is an easy exercise to reduce the σ-finite case to the finite case. We shall, hence, content ourselves with proving the above theorem when the μ_i (and, consequently, also $\mu = \mu_1 \times \mu_2$) are probability measures.

What we shall prove is that if $\mathcal{M} = \{E \in \mathcal{B} : E$ satisfies conditions (1)–(3) above$\}$, then (a) $\mathcal{A}(\mathcal{R}) \subseteq \mathcal{M}$; (b) $\mathcal{M}$ is a monotone class and (c) if we define $\mu(E)$ to be the common value of the two sides of (3) above, then μ has the properties desired of the product measure.

The proof of (a) is outlined in Ex. 4.1.6 and Ex. 4.1.7. As for (b) and (c), the verification that $\mathcal{M}$ is closed under increasing limits is a repeated application of the monotone convergence theorem. It is in proving that $\mathcal{M}$ is closed under decreasing limits that finiteness of the μ_j's is needed. Hence $\mathcal{M} \supseteq \sigma(\mathcal{R}) = \mathcal{B}_1 \otimes \mathcal{B}_2$. Finally, it only takes another easy appeal to the monotone convergence theorem to see that the common value of the two sides of the equation in the third defining property of $\mathcal{M}$, does, in fact, define a countably additive probability measure on $\mathcal{B}_1 \otimes \mathcal{B}_2$, which does all that it is claimed to in Proposition 4.1.2; the proof of that proposition is now complete.   □

Just as the 'section-approach' gave us a grasp over the product measure, a similar approach will enable us to efficiently integrate functions with respect to a product measure. We first establish the version for non-negative functions, which is called Tonelli's theorem, and then consider the general case, called Fubini's theorem.

Theorem 4.1.3   (Tonelli's theorem) *Let* $(\Omega_i, \mathcal{B}_i, \mu_i), i = 1, 2,$ *be* σ-*finite measure spaces,* $\Omega = \Omega_1 \times \Omega_2$, $\mathcal{B} = \mathcal{B}_1 \otimes \mathcal{B}_2$ *and* $\mu = \mu_1 \times \mu_2$. *Let* $f : \Omega \to \mathbb{R}$ *be a non-negative,* $(\mathcal{B}, \mathcal{B}_\mathbb{R})$-*measurable function. Then,*

(1) $f^x : \Omega_2 \to \mathbb{R}$ *is a* $(\mathcal{B}_2, \mathcal{B}_\mathbb{R})$-*measurable function for all x in* Ω_1;

(1') $f_y : \Omega_1 \to \mathbb{R}$ *is a* $(\mathcal{B}_1, \mathcal{B}_\mathbb{R})$-*measurable function for all y in* Ω_2;

(2) *the function* $x \to \int f^x(y)d\mu_2(y)$ *is a* $(\mathcal{B}_1, \mathcal{B}_{\bar{\mathbb{R}}})$-*measurable extended real-valued function on* Ω_1;

(2') *the function* $y \to \int f_y(x)d\mu_1(x)$ *is a* $(\mathcal{B}_2, \mathcal{B}_{\bar{\mathbb{R}}})$-*measurable extended real-valued function on* Ω_2;

(3) $\int_\Omega f d\mu = \int_{\Omega_1}\left(\int_{\Omega_2} f^x(y)d\mu_2(y)\right)d\mu_1(x) = \int_{\Omega_2}\left(\int_{\Omega_1} f_y(x)d\mu_1(x)\right)d\mu_2(y).$

Proof: If $f = 1_E, E \in \mathcal{B}$, the truth of this theorem is guaranteed by the discussion in the proof of Proposition 4.1.2. It follows easily that the desired conclusion holds for all non-negative simple functions.

The assertion about a general non-negative measurable function f follows now from the previous paragraph and approximation of f by non-negative simple functions—see Proposition 2.1.4—thanks to the monotone convergence theorem.   □

Remark 4.1.4 *In Tonelli's theorem, suppose, in addition, that* $f \in \mathcal{L}^1(\Omega, \mathcal{B}, \mu)$ *then* $f^x \in \mathcal{L}^1(\Omega_2, \mathcal{B}_2, \mu_2)\mu_1$-*a.e. and the function* $x \to \int f^x(y)d\mu_2(y)$ *is in* $\mathcal{L}^1(\Omega_1, \mathcal{B}_1, \mu_1)$; *an entirely similar statement holds with roles of* $(\Omega_1, \mathcal{B}_1, \mu_1)$ *and* $(\Omega_2, \mathcal{B}_2, \mu_2)$ *interchanged. This is because an integrable function is finite a.e.*

Now for the celebrated and useful Fubini's theorem.

Theorem 4.1.5 (Fubini's theorem) *Let* $(\Omega, \mathcal{B}, \mu)$ *denote the product of the* σ-*finite measure spaces* $(\Omega_1, \mathcal{B}_1, \mu_1)$ *and* $(\Omega_2, \mathcal{B}_2, \mu_2)$ *as in Theorem 4.1.2. Let* $f \in \mathcal{L}^1(\Omega, \mathcal{B}, \mu)$. *Then,*

(1) *for* μ_1-*almost all x in* Ω_1, *the function* $f^x : \Omega_2 \to \mathbb{C}$ *is* $(\mathcal{B}_2, \mathcal{B}_\mathbb{C})$-*measurable and in fact* $f^x \in \mathcal{L}^1(\Omega_2, \mathcal{B}_2, \mu_2)$;

(1') *for* μ_2-*almost all y in* Ω_2, *the function* $f_y : \Omega_1 \to \mathbb{C}$ *is* $(\mathcal{B}_1, \mathcal{B}_\mathbb{C})$-*measurable and in fact* $f_y \in \mathcal{L}^1(\Omega_1, \mathcal{B}_1, \mu_1)$;

(2) *the* μ_1-*almost everywhere defined function* $x \to \int f^x(y)d\mu_2(y)$ *is* $(\mathcal{B}_1, \mathcal{B}_\mathbb{C})$-*measurable and is in fact integrable with respect to* μ_1;

(2') *the* μ_2-*almost everywhere defined function* $y \to \int f_y(x)d\mu_1(x)$ *is* $(\mathcal{B}_2, \mathcal{B}_\mathbb{C})$-*measurable and is in fact integrable with respect to* μ_2; *and*

(3)

$$\int_\Omega f d\mu = \int_{\Omega_1}\left(\int_{\Omega_2} f(x, y)d\mu_2(y)\right)d\mu_1(x)$$

$$= \int_{\Omega_2}\left(\int_{\Omega_1} f(x, y)d\mu_1(x)\right)d\mu_2(y).$$

Proof: For non-negative f, Tonelli's theorem implies Fubini's theorem. A general f admits a canonical decomposition $f = f_1 - f_2 + i(f_3 - f_4)$ in terms of non-negative functions in such a way that f is integrable if and only if each f_i is. This, together with Remark 4.1.4 implies Fubini's theorem.   □

EXERCISES

Ex. 4.1.6   Let $\{A_1, \dots, A_n\}$ be subsets of any set Ω. The aim of this exercise is to explicitly describe the algebra $\mathcal{A}(\{A_1, \dots, A_n\})$, and in the process show that it is a finite collection whose cardinality is a power of 2.

For $\epsilon = (\epsilon_1, \dots, \epsilon_n) \in \{0, 1\}^n$, define

$$A(\epsilon) = \cap_{i=1}^n A_i^{(\epsilon_i)}, \text{ where we write } B^{(i)} = \begin{cases} B & if \ i = 1 \\ B' & if \ i = 0. \end{cases}$$

Show that

(1) $\Omega = \bigsqcup_{\epsilon \in \{0,1\}^n} A(\epsilon)$;
(2) $A_i = \cup\{A(\epsilon) : \epsilon(i) = 1\}$;
(3) $\{A(\epsilon) : A(\epsilon) \neq \emptyset\}$ is the collection of 'atoms' (= minimal non-empty elements) of $\mathcal{A}(\{A_1, \dots, A_n\})$ and
(4)   if $|\{\epsilon : A(\epsilon) \neq \emptyset\}| = k$, then $\mathcal{A}(\{A_1, \dots, A_n\})$ has 2^k elements.

Ex. 4.1.7   Suppose $E = \bigsqcup_{i=1}^n (A_i \times B_i)$. Let $\{\alpha_1, \dots, \alpha_p\}$ be the set of atoms of $\mathcal{A}(\{A_1, \dots, A_n\})$ as in Ex. 4.1.6 above. Similarly, let $\{\beta_1, \dots, \beta_q\}$ be the set of atoms of $\mathcal{A}(\{B_1, \dots, B_n\})$. Then, show that

(1) $\{\alpha_i \times \beta_j : 1 \le p \le m, 1 \le q \le n\}$ is the set of atoms of $\mathcal{A}(\{A_i \times B_j : 1 \le i, j \le n\})$;
(2)   there exists a set $S \subset \{1, \dots, p\} \times \{1, \dots, q\}$ such that $E = \bigsqcup_{\{(i,j) \in S\}} \alpha_i \times \beta_j$;
(3) $E^x = \bigsqcup_{j \in S^i} \beta_j$, for $x \in \alpha_i$, where of course $S^i = \{1 \le j \le q : (i, j) \in S\}$ and $\mu_2(E^x) = \sum_{i=1}^p \left(\sum_{j \in S^i} \mu_2(\beta_j) \right) 1_{A_i}(x)$;
(4) $E_y = \bigsqcup_{i \in S_j} \alpha_i$, for $y \in \beta_j$, where of course $S_j = \{1 \le i \le p : (i, j) \in S\}$ and $\mu_1(E_y) = \sum_{j=1}^q \left(\sum_{i \in S_j} \mu_1(\alpha_i) \right) 1_{\beta_j}(y)$ and
(5)

$$\int \mu_2(E^x) d\mu_1(x) = \sum_{i=1}^p \left(\sum_{j \in S^i} \mu_2(\beta_j) \right) \mu_1(\alpha_i)$$

$$= \sum_{(i,j) \in S} \mu_1(\alpha_i) \mu_2(\beta_j)$$

$$= \sum_{j=1}^{q} \left(\sum_{i \in S_j} \mu_1(\alpha_i) \right) \mu_2(\beta_j)$$

$$= \int \mu_1(E_y) d\mu_2(y);$$

(6) $A(\mathcal{R}) \subset \mathcal{M}$.

(It must be mentioned that the equality of the expressions at the two ends of the equation array in (5) can also be proved directly; the reader is urged to try and discover this direct verification.)

Ex. 4.1.8   Suppose $(\Omega, \mathcal{B}, \mu)$ is a σ-finite measure space and $f : \Omega \to \mathbb{R}$ is a non-negative $(\mathcal{B}, \mathcal{B}_{\mathbb{R}})$-measurable function. Define $\mathcal{G} = \{(w, t) \in \Omega \times \mathbb{R} : 0 \le t \le f(w)\}$. Show that $\mathcal{G} \in \mathcal{B} \otimes \mathcal{B}_{\mathbb{R}}$, and that $(\mu \times m)(\mathcal{G}) = \int f d\mu$, where m denotes the Lebesgue measure on $\mathbb{R}$. (*Hint:* First assume f is simple; then use simple approximation.)

Ex. 4.1.9   Extend the notion of product measure to $\prod_{i=1}^{n} (\Omega_i, \mathcal{B}_i, \mu_i)$.

Ex. 4.1.10   Let $\Omega_1 = \Omega_2 = \mathbb{N}$, $\mathcal{B}_1 = \mathcal{B}_2 = \mathcal{P}(\mathbb{N})$. Define the measures μ_i on $(\Omega_i, \mathcal{B}_i)$ by $\mu_i(\{k\}) = 2^{-k}$. Define $f : \Omega_1 \times \Omega_2 \to \mathbb{R}$ by

$$f(m, n) = \begin{cases} -n2^{2n} & \text{if } m = n \\ n2^{2n} & \text{if } m = n - 1 \\ 0 & \text{otherwise} \end{cases}$$

(a)   Show that f is $\mathcal{B}_1 \otimes \mathcal{B}_2$-measurable.
(b)   Observe that

$$\int_{\Omega_2} \int_{\Omega_1} f(m, n) d\mu_1(m) d\mu_2(n) \neq \int_{\Omega_1} \int_{\Omega_2} f(m, n) d\mu_2(n) d\mu_1(m),$$

thereby emphasising the importance of integrability in the hypotheses of Fubini's theorem.

Ex. 4.1.11   Consider the σ-algebra $\mathcal{C}$ whose members are subsets of $\mathbb{R}$ which are either countable or have countable complements and show that the diagonal $D = \{(x, x) : x \in \mathbb{R}\}$ does not belong to $\mathcal{C} \otimes \mathcal{C}$ although all its sections D^x and D_y belong to $\mathcal{C}$.

4.2 Joint Distribution and Independence

In this section, we relate the notions of independent random variables and product measures. The link is via the notion of joint distribution.

Definition 4.2.1 *Let X and Y be two real-valued random variables defined on $(\Omega, \mathcal{B}, P)$. The **joint distribution** of X, Y is the probability measure $P_{(X,Y)}$ on $(\mathbb{R}^2, \mathcal{B}_{\mathbb{R}^2})$ defined by*

$$P_{(X,Y)}(E) = P((X, Y) \in E) \ \forall E \in \mathcal{B}_{\mathbb{R}^2}.$$

Proposition 4.2.2 *Suppose X, Y are two random variables on the probability space $(\Omega, \mathcal{B}, P)$. Then X, Y are independent if and only if $P_{(X,Y)} = P_X \times P_Y$ (i.e. the joint distribution is the product of the individual distributions).*

Proof: By definition, X and Y are independent if and only if $P(X \in A, Y \in B) = P(X \in A)P(Y \in B) \forall A, B \in \mathcal{B}_{\mathbb{R}}$, which holds if and only if $P_{(X,Y)}(A \times B) = P(X \in A)P(Y \in B) \forall A, B \in \mathcal{B}_{\mathbb{R}}$. By Proposition 4.1.2, we are done. □

Suppose X, Y are random variables on the probability space $(\Omega, \mathcal{B}, P)$. Let $Z = X + Y$. The distribution of Z when X and Y are independent has a special form. Observe that $P(Z \leq a) = P(X + Y \leq a) = P_{(X,Y)}(B_a)$ where $B_a = \{(x, y) : x + y \leq a\}$. Now, $P_{(X,Y)}(B_a) = P_X \times P_Y(B_a) = \int P_Y(B_a^x) d P_X(x)$. Hence, we have

$$F_z(a) = \int F_Y(a - x) d P_X(x).$$

If, in addition, X and Y have densities f_X and f_Y then, from the above, we see that so does Z and it is given by $f_Z(z) = \int f_Y(z - x) f_X(x) dx$.

EXERCISES

Ex. 4.2.3 Let X, Y be independent random variables on a probability space $(\Omega, \mathcal{B}, P)$ such that $E(X)$ and $E(Y)$ exist. Then show that $E(XY) = E(X)E(Y)$. Is the converse true?

Ex. 4.2.4 Let X be random variable on a probability space $(\Omega, \mathcal{B}, P)$. Then, show that X is independent of X if and only if X is a degenerate random variable (i.e., there is a $c \in \mathbb{R}$ such that $X = c$ almost everywhere).

Ex. 4.2.5 If X, Y are random variables on a probability space $(\Omega, \mathcal{B}, P)$ with joint distribution $P_{(X,Y)}$, consider the random variables $\tilde{X}, \tilde{Y}$ defined on the probability space $(\mathbb{R}^2, \mathcal{B}_{\mathbb{R}^2}, P_{(X,Y)})$ by $\widetilde{X}(x, y) = x, \widetilde{Y}(x, y) = y$. Show that $P_{(\widetilde{X},\widetilde{Y})} = P_{(X,Y)}$.

4.3 Probability Measures on Infinite Product Spaces

Let $(\Omega_i, \mathcal{B}_i)$, $i \in I$ be any collection of measurable spaces. Let Ω denote the 'product space' $\Omega = \prod_{i \in I} \Omega_i$. (Thus, an element of Ω may be thought of as a 'tuple' $w = (w_i)_{i \in I}$ with $w_i \in \Omega_i \ \forall i \in I$; more formally, an element of Ω may be thought of as a function $w : I \to \cup_{i \in I} \Omega_i$ such that $w(i) \in \Omega_i \ \forall i \in I$.)

If I is finite and some specific ordering of I is chosen, say $I = \{i_1, \dots, i_n\}$, then $\Omega = \prod_{i \in I} \Omega_i$ may be identified with the usual Cartesian product, $\Omega_{i_1} \times \cdots \times \Omega_{i_n}$.

Even if I is infinite, consider the 'finite products' $\Omega_F = \prod_{i \in F} \Omega_i$, as F ranges over finite subsets of I. There is an obvious surjective 'projection' mapping, $\pi_F : \Omega \to \Omega_F$ (obtained by 'viewing only the co-ordinates coming from F'). When $F = \{i\}$ is a singleton, we identify $\Omega_{\{i\}}$ with Ω_i and write π_i for $\pi_{\{i\}}$: thus $\pi_i((w_j)_{j \in I}) = w_i$.

With the foregoing notation, we define $\otimes_{i \in I} \mathcal{B}_i$ to be the smallest σ-algebra $\mathcal{B}$ on $\Omega = \prod_{i \in I} \Omega_i$ with the property that the coordinate projections $\pi_i, i \in I$ are $(\mathcal{B}, \mathcal{B}_i)$-measurable for all $i \in I$; thus, $\mathcal{B} = \sigma(\{\pi_i^{-1}(E_i) : E_i \in \mathcal{B}_i, i \in I\})$.

Ex. 4.3.1

(1) If I is finite, say $I = \{1, 2, \dots, n\}$ and $(\Omega_i, \mathcal{B}_i), i = 1, 2, \dots, n$ are measurable spaces, show that $\otimes_{i \in I} \mathcal{B}_i = \sigma(\{E_1 \times \cdots \times E_n : E_i \in \mathcal{B}_i \ \forall i\})$.

(2) If I is infinite, and $F \subseteq I$ is a finite set, show that $\pi_F : \Omega \ (= \prod_{i \in I} \Omega_i) \to \Omega_F = \prod_{i \in F} \Omega_i$ is $(\otimes_{i \in I} \mathcal{B}_i, \otimes_{i \in F} \mathcal{B}_i)$-measurable.

If $\Omega = \prod_{i \in I} \Omega_i$, $\mathcal{B} = \otimes_{i \in I} \mathcal{B}_i$ as above, and if $F \subseteq I$ is a finite set of indices, define $\mathcal{B}_F = \{\pi_F^{-1}(E) : E \in \otimes_{i \in F} \mathcal{B}_i\}$. An element of $\mathcal{B}_F$, for F any finite subset of I, is called a 'finite-dimensional cylinder set' in Ω. Thus, $\pi_F^{-1}(E)$ is thought of as a 'cylinder on base E'. Clearly, $\mathcal{B}_F$ is a σ-algebra of subsets of Ω and the mapping $E \mapsto \pi_F^{-1}(E)$ defines an isomorphism (i.e., a bijection preserving countable unions and complements) of the σ-algebra $\otimes_{i \in F} \mathcal{B}_i$ onto $\mathcal{B}_F$.

If F and G are finite subsets of I such that $F \subseteq G$, there is a natural projection $\pi_F^G : \Omega_G \to \Omega_F$ such that $\pi_F = \pi_F^G \circ \pi_G$. Hence, $\mathcal{B}_F \subseteq \mathcal{B}_G$.

Note, next, that the collection $\mathcal{F}_I$ of finite subsets of I forms a directed set under inclusion, i.e., $\mathcal{F}_I$ is partially ordered : $F_1 \leq F_2$ iff $F_1 \subseteq F_2$; and if $F_1, F_2 \in \mathcal{F}_I$, then there exists $F_0 (= F_1 \cup F_2) \in \mathcal{F}_I$ such that $F_1 \leq F_0$ and $F_2 \leq F_0$.

It follows that $\{\mathcal{B}_F : F \in \mathcal{F}_1\}$ is a 'net' of σ-algebras of subsets of Ω, i.e., whenever $F_1, F_2 \in \mathcal{F}_1$, there exists $F_0 \in \mathcal{F}_1$ such that $\mathcal{B}_{F_1} \cup \mathcal{B}_{F_2} \subseteq \mathcal{B}_{F_0}$ (so that $\sigma(\mathcal{B}_{F_1} \cup \mathcal{B}_{F_2}) \subseteq \mathcal{B}_{F_0}$).

The preceding analysis shows that $\mathcal{A} = \cup_{F \in \mathcal{F}_1} \mathcal{B}_F$ is an algebra of subsets of Ω and that the product σ-algebra $\mathcal{B} = \otimes_{i \in I} \mathcal{B}_i$ is, more or less by definition, given by $\mathcal{B} = \sigma(\mathcal{A})$.

Hence, in view of Caratheodory's extension theorem, in order to construct measures on $(\Omega, \mathcal{B})$, it is sufficient to construct measures on the algebra $\mathcal{A}$.

Since $\mathcal{A} = \cup_{F \in \mathcal{F}_1} \mathcal{B}_F$, in order to construct a measure μ on $\mathcal{A}$ we need to construct a measure μ_F on $\mathcal{B}_F$ (where $\mu_F = \mu|\mathcal{B}_F$) for each $F \in \mathcal{F}_1$ subject to the following two consistency conditions:

(1) $F_1, F_2 \in \mathcal{F}_1, F_1 \subset F_2 \Rightarrow \mu_{F_2}|\mathcal{B}_{F_1} = \mu_{F_1}$;
(2)   if $A = \coprod_{n=1}^{\infty} A_n$, with $A \in \mathcal{B}_F$, $A_n \in \mathcal{B}_{F_n}$, for $F, F_1, F_2, \dots, \in \mathcal{F}_1$, then
$\mu_F(A) = \sum_{n=1}^{\infty} \mu_{F_n}(A_n)$.

(Given such a family, simply define μ on $\mathcal{A}$ by the prescription $\mu|\mathcal{B}_F = \mu_F$; this is well-defined by (1), and countably additive by (2)).

Ex. 4.3.2   Let $(\Omega_i, \mathcal{B}_i)$, $i \in I$, be measurable spaces and let $\Omega = \prod_{i \in I} \Omega_i$, $\mathcal{B} = \otimes_{i \in I} \mathcal{B}_i$. If $(\Omega', \mathcal{B}')$ is a measurable space, show that a map $T : \Omega' \to \Omega$ is $(\mathcal{B}', \mathcal{B})$-measurable if and only if $\pi_i \circ T$ is $(\mathcal{B}', \mathcal{B}_i)$-measurable for each $i \in I$.

Remark 4.3.3   *Suppose* $\{X_i : i \in I\}$ *is a collection of random variables defined on a probability space* $(\Omega_0, \mathcal{B}_0, P_0)$. *Let* $\Omega = \mathbb{R}^I, \mathcal{B} = \otimes_{i \in I} \mathcal{B}_{\mathbb{R}}$. *(Thus, if* $(\Omega_i, \mathcal{B}_i) = (\mathbb{R}, \mathcal{B}_{\mathbb{R}})$ *for each i, then* $\Omega = \prod_{i \in I} \Omega_i$ *and* $\mathcal{B} = \otimes_{i \in I} \mathcal{B}_i$.*) Define* $T : \Omega_0 \to \Omega$ *by* $T(w) = ((X_i(w)))_{i \in I}$. *Then* T *is* $(\mathcal{B}_0, \mathcal{B})$-*measurable and* $T P_0 (= P_0 \circ T^{-1})$ *is a probability measure defined on* $(\Omega, \mathcal{B})$ *(and may be called the joint distribution of the collection* $\{X_i : i \in I\}$ *of random variables).*

Rather than getting caught up in extra generality which can easily be treated once we have understood the special case we shall look at, we shall, in the rest of this section, consider the special case of the infinite product where $I = \mathbb{N} = \{1, 2, 3, \dots\}$ and all the factors $(\Omega_i, \mathcal{B}_i)$ are the same measurable space $(\Omega_0, \mathcal{B}_0)$.

Thus, we shall assume that $(\Omega_0, \mathcal{B}_0)$ is a fixed measurable space, and that $\Omega = \Omega_0^{\mathbb{N}} = \{w = ((w_n))_{n=1}^{\infty} : w_n \in \Omega_0 \ \forall n\}$ and $\mathcal{B} = \otimes_{n=1}^{\infty} \mathcal{B}_0 = \sigma(\mathcal{A})$, where $\mathcal{A}$ is the algebra of all finite-dimensional cylinder sets in Ω, as discussed earlier.

We shall employ the following notation: for $n \in \mathbb{N}$, $\pi_n : \Omega \to \Omega_0$ will denote the map $\pi_n((w_k)) = w_n$, while $\pi_{n]}$ will denote the projection mapping $\pi_{n]} : \Omega \to \Omega_0^n$ given by $\pi_{n]}(w) = (w_1, w_2, \dots, w_n)$; also, we let $\mathcal{B}_{n]}$ denote the σ-algebra of subsets of Ω defined by $\mathcal{B}_{n]} = \{\pi_{n]}^{-1}(E) : E \in \otimes_{k=1}^{n} \mathcal{B}_0\}$. Thus, $\mathcal{A} = \cup_{n=1}^{\infty} \mathcal{B}_{n]}$.

Our goal is to construct several 'interesting' probability measures on $(\Omega, \mathcal{B})$.

Such a probability measure may be thought of (at least when $\Omega_0 \subseteq \mathbb{R}$)—cf. Remark 4.3.3—as the joint distribution of a sequence of random variables $\{X_n\}_{n=1}^{\infty}$, where each X_n takes values in the 'state space Ω_0'. Such a sequence $\{X_n\}_{n=1}^{\infty}$ is sometimes called a 'discrete-time stochastic process taking values in Ω_0'. (The word 'discrete' is used here, since one also sometimes considers 'continuous-time stochastic processes', i.e., collections $\{X_t : t \in [a, b]\}$ of random variables indexed by an interval in the line; we shall briefly touch on this aspect later in the notes.)

To make matters as simple as possible, we shall, in the rest of this section, consider the case where the state space Ω_0 is a finite set, and then consider more general cases in the next section.

We pause to examine the motivating example of the 'random walk on the vertices of an N-gon'. Thus, let Ω_0 denote the set of N distinct vertices $w_1, \ldots, w_N$ of an N-gon. Suppose that at time 0, a marker is placed on vertex w_0. After one unit of time, toss a fair coin and then move the marker to the vertex that is immediately to the left or right of the current vertex according to whether the toss results in heads or tails, i.e., the marker moves one step to the left or right with equal probability, 1/2. After one more unit of time, repeat this process and keep iterating this procedure *ad infinitum*. Let X_n denote the vertex at which the marker finds itself after n units of time. Then $\{X_n\}_{n=1}^{\infty}$ is an example of a discrete-time stochastic process on the finite state space Ω_0. To treat such an example, we would like to construct a suitable probability space $(\Omega, \mathcal{B}, P)$ and random variables $X_n : \Omega \rightarrow \{1, \ldots, N\}$ such that $P(X_{n_1} = j_1, \ldots, X_{n_k} = j_k)$ is precisely the probability that the marker is at the vertices $w_{j_1}, \ldots, w_{j_k}$ at times $n_1, \ldots, n_k$, respectively.

In the language of Remark 4.3.3, the joint distribution of the sequence $\{X_n\}_{n=1}^{\infty}$ would be a probability measure P defined on $(\Omega, \mathcal{B})$, where $\Omega = \Omega_0^{\mathbb{N}}$ and $\mathcal{B} = \otimes_{n=1}^{\infty} \mathcal{B}_0$, where $\mathcal{B}_0 = 2^{\Omega_0}$ is the σ-algebra of all subsets of Ω_0, and we may consider the random variables $\{X_n\}_{n=1}^{\infty}$ defined on $(\Omega, \mathcal{B}, P)$ by $X_n((w_k)_{k=1}^{\infty}) = w_n$.

In the rest of this section, we assume that $\Omega_0 = \{1, 2, \ldots, N\}$, $\mathcal{B}_0 = 2^{\Omega_0}$ and $\Omega = \Omega_0^{\mathbb{N}}$ (is the set of infinite sequences in Ω_0) and $\mathcal{B} = \otimes_{n=1}^{\infty} \mathcal{B}_0$. Further, we write $X_n : \Omega \rightarrow \Omega_0$ for the nth coordinate projection, for $n = 1, 2, \ldots$. It must be clear that Ω_0^n is finite and that $\otimes_{k=1}^{n} \mathcal{B}_0 (= 2^{\Omega_0^n})$ consists of all subsets of Ω_0^n. A typical n-dimensional cylinder set, i.e., an element of $\mathcal{B}_{n]}$, in the notation introduced earlier, is just a set of the form $\{w \in \Omega : (X_1(w), \ldots, X_n(w)) \in E\}$ where $E \subseteq \Omega_0^n$ is any subset. Thus a set in $\mathcal{B}_{n]}$ is described by a statement about the trajectory $(X_1(w), X_2(w), \ldots)$ up to time n.

Before proceeding to the main result, it will be convenient to single out an elementary fact as a lemma.

Lemma 4.3.4   *Suppose $\mathcal{A}$ is an algebra of subsets of a set Ω. Suppose μ : $\mathcal{A} \to [0, 1]$ is a finitely additive set function. The following conditions on μ are equivalent:*

(1)   *μ is countably additive on $\mathcal{A}$;*
(2)   *μ is 'continuous from above at the empty set': i.e., if $A_n \in \mathcal{A}$, $A_1 \supseteq A_2 \supseteq \dots$ and $\cap_{n=1}^{\infty} A_n = \phi$, then $\lim_{n \to \infty} \mu(A_n) = 0$.*

Proof:

(1) $\Rightarrow$ (2) since all finite measures are continuous from above.
(2) $\Rightarrow$ (1) : suppose $B, B_1, B_2, \dots, \in \mathcal{A}$ and $B = \coprod_{n=1}^{\infty} B_n$. Let $S_n = \coprod_{k=1}^{n} B_k$. Then, $S_n \in \mathcal{A}$ since $\mathcal{A}$ is an algebra, and $\mu(S_n) = \sum_{k=1}^{n} \mu(B_k)$ since μ is finitely additive.

Now, if $A_n = B - S_n$, then $A_n \in \mathcal{A}$; also $A_1 \supseteq A_2 \supseteq \dots$ and $\cap A_n = \phi$. Hence, $\lim_{n \to \infty} \mu(A_n) = 0$. But $\mu(B) = \mu(S_n) + \mu(A_n)$ $\forall n$ (by finite additivity), and hence $\mu(B) = \lim_{n \to \infty} \mu(S_n) = \sum_{k=1}^{\infty} \mu(B_k)$.   □

Theorem 4.3.5   *Let $(\Omega, \mathcal{B})$ be as above.*

(1)   *Let P be a probability measure defined on $\mathcal{B}$. Define*

$$p_{i_1,\dots,i_n} = P(\{w \in \Omega : X_1(w) = i_1, \dots, X_n(w) = i_n\}) \qquad (4.3.1)$$

for arbitrary $n = 1, 2, \dots$ and $i_1, \dots, i_n \in \{1, 2, \dots, N\}$. Then, for all $i_1, \dots, i_n \in \{1, 2, \dots, N\}$ and for all $n = 1, 2, \dots,$

$$p_{i_1,\dots,i_n} \geq 0;$$

$$\sum_{i_1,\dots,i_n=1}^{N} p_{i_1,\dots,i_n} = 1; \qquad (4.3.2)$$

$$p_{i_1,\dots,i_n} = \sum_{j=1}^{N} p_{i_1,\dots,i_n,j}.$$

(2)   *Conversely, if $\{p_{i_1,\dots,i_n} : n = 1, 2, \dots, i_1, \dots, i_n \in \{1, \dots, N\}\}$ are arbitrary numbers satisfying (4.3.2), then there exists a unique probability measure, P, on $(\Omega, \mathcal{B})$ satisfying (4.3.1).*

Proof:

(1)   Suppose P is a probability measure on $(\Omega, \mathcal{B})$ and $p_{i_1,\dots,i_n} = P(\pi_n^{-1}(\{(i_1, \dots, i_n)\}))$. Clearly, $p_{i_1,\dots,i_n} \geq 0$, and $\sum_{i_1,\dots,i_n=1}^{N} p_{i_1,\dots,i_n} = 1$, since $\Omega = \coprod_{i_1,\dots,i_n=1}^{N} \pi_n^{-1}(\{(i_1, \dots, i_n)\})$. Also, since $\pi_n^{-1}(\{(i_1, \dots, i_n)\}) = \coprod_{j=1}^{N} \pi_{n+1}^{-1}(\{(i_1, \dots, i_n, j)\})$, it is clear that $p_{i_1,\dots,i_n} = \sum_{j=1}^{N} p_{i_1,\dots,i_n,j}$.

(2)   Suppose we are given constants $p_{i_1,\ldots,i_n}$ satisfying (4.3.2). For each n, define
$\mu_n : \mathcal{B}_{n]} \to [0, 1]$ by $\sum_{i_1,\ldots,i_n=1}^{N} 1_E(i_1, \ldots, i_n) p_{i_1,\ldots,i_n} = \mu_n(\pi_n^{-1}(E))$.
The hypotheses on the $p_{i_1,\ldots,i_n}$'s ensure that μ_n is a probability measure
defined on $\mathcal{B}_{n]}$, and that $\mu_{n+1}|\mathcal{B}_{n]} = \mu_n$.

So, we may unambiguously define $P : \mathcal{A} = (\cup_{n=1}^{\infty} \mathcal{B}_{n]}) \to [0, 1]$ by $P|\mathcal{B}_{n]} = \mu_n$.
It is clear that P is a finitely additive set function. In order to prove the theorem,
it suffices, thanks to the Caratheodory extension theorem, to verify that P is
countably additive on the algebra $\mathcal{A}$ since $\mathcal{B} = \sigma(\mathcal{A})$.
    In view of Lemma 4.3.4, we need to check that P is continuous from above
at the empty set.
    We make the even-stronger assertion that if $A_n \in \mathcal{A}$, $A_1 \supseteq A_2 \supseteq \cdots$ and
$\cap A_n = \phi$, then $A_n = \phi$ for large n. (Hence any finitely additive set function in
$\mathcal{A}$ which assigns mass 1 to Ω, is necessarily countably additive.) Equivalently,
we need to show that if $A_n \in \mathcal{A}$, $A_1 \supseteq A_2 \supseteq \cdots$ and $A_n \neq \phi$ for all n, then
$\cap A_n \neq \phi$.
    One (slick) proof of this comes from the kind of topological considerations
discussed briefly in the section on Bernoulli trials. The reader is encouraged
to furnish a 'bare-hands' proof of the fact using the observation that if A_n are
as above, then, for each k $\{\pi_k(A_n)\}_{n=1}^{\infty}$ is a decreasing sequence of non-empty
subsets of the finite set $\{1, 2, \ldots, N\}$.   □

We single out a special, but very important, case below. Markov chains are
discussed in some detail in Chapter 6.

Definition 4.3.6   *Let $\Omega, \mathcal{B}, P, p_{i_1,\ldots,i_n}$ be as in Theorem 4.3.5. We say that
the associated sequence $\{X_n\}_{n=1}^{\infty}$ of coordinate projections defines a finite-state
Markov chain if there exists a sequence of $N \times N$ matrices $\pi^{(n)} = ((\pi_{ij}^{(n)}))$ such
that $p_{i_1,\ldots,i_{n+1}} = p_{i_1,\ldots,i_n} \pi_{i_n i_{n+1}}^{(n)}$ for all $i_1, \ldots, i_{n+1}$ and for all $n \geq 1$. The matrices
$((\pi_{ij}^{(n)}))$ are called the transition probability matrices of the Markov chain.*

    If $\{X_n\}_{n=1}^{\infty}$ is a finite state Markov chain, and $p_{i_1,\ldots,i_n}$ and $\pi_{ij}^{(n)}$ are as above,
note that

$$p_{i_1,\ldots,i_n} = p_{i_1} \pi_{i_1 i_2}^{(1)} \pi_{i_2 i_3}^{(2)} \cdots \pi_{i_{n-1} i_n}^{(n-1)} \tag{4.3.3}$$

for all $i_1, \ldots, i_n$, for all n. The quantity $\pi_{ij}^{(n)}$ is to be thought of as the probability
of transition from the state i to the state j at time n. Note that the third condition
in (4.3.2) implies that $\sum_{j=1}^{N} \pi_{ij}^{(n)} = 1$ $\forall n = 1, 2, \ldots$ and for all $i = 1, \ldots, N$,
while the second condition of (4.3.2) implies that $\sum_{i=1}^{n} p_i = 1$.
    A square matrix, (π_{ij}), with non-negative entries is called a **stochastic matrix**
if $\sum_j \pi_{ij} = 1$ for all i.

Conversely, if we are given any sequence $\{\pi^{(n)}\} = \{((\pi_{ij}^{(n)}))\}_{n=1}^{\infty}$ of $N \times N$ stochastic matrices, and non-negative numbers $p_1, \ldots, p_N$ such that $\sum_{i=1}^{n} p_i = 1$, it must be clear that if we define $p_{i_1,\ldots,i_n} = p_{i_1} \pi_{i_1 i_2}^{(n)} \cdots \pi_{i_1 i_n}^{(n-1)}$, then the $p_{i_1,\ldots,i_n}$'s satisfy (4.3.2), and hence give rise to a Markov chain.

Thus, any Markov chain is completely described by (i) the sequence $\{\pi^{(n)}\}_{n=1}^{\infty}$ of transition probability matrices and (ii) the 'initial distribution' $p_1, \ldots, p_N$. (Note that $p_j = P(X_1 = j)$.)

A Markov chain is said to be stationary if

(1) $\pi^{(1)} = \pi^{(2)} = \cdots = \pi$ (say), and
(2) $\sum_{i=1}^{N} p_i \pi_{ij} = p_j$ for all j, or, in the notation of matrix multiplication,

$$[p_1 \ldots p_N] \begin{bmatrix} \pi_{11} & \cdots & \pi_{1N} \\ \vdots & & \vdots \\ \pi_{N1} & \cdots & \pi_{NN} \end{bmatrix} \doteq [p_1 \ldots p_N].$$

Ex. 4.3.7   Suppose $\{X_n\}$ is a stationary Markov chain in the sense mentioned above.

(1)   For any two positive integers m and n, show that $(X_1, \ldots, X_m)$ and $(X_{n+1}, \ldots, X_{n+m})$ have the same joint distribution.
(2)   Deduce from (1) that for any positive integer n, the sequences $(X_1, X_2, \ldots)$ and $(X_{n+1}, X_{n+2}, \ldots)$ have the same joint distributions in the sense of Remark 4.3.3. (This is the reason for the use of the word 'stationary'.)

Ex. 4.3.8   Show that the random walk on the vertices of an N-gon that was discussed earlier is a Markov chain; what are the transition probabilities?

Ex. 4.3.9   Formulate and prove a modified version of Theorem 4.3.5 which applies to $\Omega = \prod_{i=1}^{\infty} \Omega_i$, where the $\Omega_i's$ are finite and not necessarily of the same size.

4.4   Kolmogorov Consistency Theorem

We assume throughout this section that $\Omega = \mathbb{R}^I$ and $\mathcal{B} = \otimes_{i \in I} \mathcal{B}_{\mathbb{R}}$, where I is an arbitrary index set. (Thus $\Omega = \prod_{i \in I} \Omega_i$ and $\mathcal{B} = \otimes_{i \in I} \mathcal{B}_i$, where $(\Omega_i, \mathcal{B}_i) = (\mathbb{R}, \mathcal{B}_{\mathbb{R}})$ for all i in I.) We continue to use the notation of the last section. Thus $\mathcal{F}_I$ denotes the family of finite subsets of I; if $F \in \mathcal{F}_I$, we write $\pi_F : \Omega \rightarrow \Omega_F = \mathbb{R}^F$ for the natural projection, and let $\mathcal{B}_F = \{\pi_F^{-1}(E) : E \in \otimes_{i \in F} \mathcal{B}_{\mathbb{R}}\}$. (Thus, the map $E \rightarrow \pi_F^{-1}(E)$ is an isomorphism of the σ-algebra $\otimes_{i \in F} \mathcal{B}_{\mathbb{R}}$ onto $\mathcal{B}_F$, and the map $\mu \rightarrow \mu \circ \pi_F^{-1}$ sets up a bijection between the collections of probability measures on $\mathcal{B}_F$ and on $\otimes_{i \in F} \mathcal{B}_{\mathbb{R}}$—see Ex. 1.8.3.)

Recall that, in order to define a probability measure P on $(\Omega\ \mathcal{B})$, it is necessary and sufficient that we construct a family $\{P_F\ :\ F\ \in\ \mathcal{F}_I\}$, where P_F is a probability measure defined on $(\Omega\ ,\mathcal{B}_F)$ so that the family $\{P_F\ :\ F\ \in\ \mathcal{F}_I\}$ satisfies two conditions:

(1)   (Consistency) if $F, G \in \mathcal{F}_I$ and $F \subseteq G$, then $P_G|\mathcal{B}_F = P_F$; and
(2)   (Countable additivity) if $A_n \in \mathcal{B}_{F_n}$, $F_1, F_2, \ldots , \in \mathcal{F}_I$ and if $A_1 \supseteq A_2 \supseteq \ldots \searrow \phi$, then $\lim_{n\to\infty} P_{F_n}(A_n) = 0$.

The main result in this section is 'Kolmogorov's consistency theorem' which says that if the consistency condition above is satisfied, then countable additivity is a consequence of the fact that all probability measures defined on $(\mathbb{R}^n, \mathcal{B}_{\mathbb{R}^n})$ are automatically *regular*. We prove this fact concerning regularity of probability measures on $(\mathbb{R}^n, \mathcal{B}_{\mathbb{R}^n})$ for the case when $n = 1$, and leave the general case as an exercise.

Lemma 4.4.1   *Let P be a probability measure defined on $(\mathbb{R}, \mathcal{B}_{\mathbb{R}})$. Then, for all $E \in \mathcal{B}_{\mathbb{R}}$ and for all $\varepsilon > 0$, there exists an open set U such that $E \subseteq U$ and $P(U) < P(E) + \varepsilon$.*

Proof: Recall that if $\mathcal{A}$ is the collection of sets of the form $A = \prod_{i=1}^{n}[a_i, b_i)$, with $-\infty \leq a_1 < b_1 < a_2 < \cdots < a_n < b_n \leq \infty$ (with the convention that $[-\infty, b)$ is to be interpreted as $(-\infty, b)$), then $\mathcal{A}$ is an algebra of subsets of $\mathbb{R}$ and $\mathcal{B}_{\mathbb{R}} = \sigma (\mathcal{A})$. Since $\mu(E) = \mu^*(E)$ for all $E \in \mathcal{B}_{\mathbb{R}}$, it is clear from the definition of μ^* (and of $\mathcal{A}$) that there exist intervals $[a_n, b_n)$, $-\infty \leq a_n < b_n \leq \infty$ such that $E \subseteq \bigcup_{n=1}^{\infty}[a_n, b_n)$ and $\sum_{n=1}^{\infty} P([a_n, b_n)) < P(E) + \varepsilon/2$. It is clear that we may find $\varepsilon_n > 0$ so that $P((a_n - \varepsilon_n, b_n)) < P([a_n, b_n)) + \varepsilon/2^{n+1}$. Put $U = \bigcup_{n=1}^{\infty}(a_n - \varepsilon_n, b_n)$ and note that $P(U) \leq \sum_{n=1}^{\infty} P((a_n - \varepsilon_n, b_n)) < P(E) + \varepsilon$. $\square$

Recall that a subset of $\mathbb{R}$ is compact if it is closed and bounded. We are now ready to establish that probability measures on $(\mathbb{R}\ , \mathcal{B}_{\mathbb{R}})$ are *inner regular*.

Proposition 4.4.2   *If P is a probability measure defined on $(\mathbb{R}, \mathcal{B}_{\mathbb{R}})$, and if $E \in \mathcal{B}_{\mathbb{R}}$, then for all $\varepsilon > 0$, there exists a compact set $K \subseteq E$ such that $P(K) > P(E) - \varepsilon$.*

Proof: Apply Lemma 4.4.1 to E' to find an open set $U \supseteq E'$ such that $P(U) < P(E')+\varepsilon/2$. It follows that $F = U'$ is a closed set, that $F \subseteq E$ and that $P(F) > P(E) - \varepsilon/2$. If $K_n = F \cap [-n, n]$, then $K_n \nearrow F$ and so $\lim_{n\to\infty} P(K_n) = P(F)$. Set $K = K_n$ for any sufficiently large n. $\square$

Ex. 4.4.3   Show that every probability measure P defined on $(\mathbb{R}^n, \mathcal{B}_{\mathbb{R}^n})$ is regular, i.e., if $E \in \mathcal{B}_{\mathbb{R}^n}$ and $\varepsilon > 0$, then there exist an open set U and a compact set K such that $K \subseteq E \subseteq U$ and $P(U - K) < \varepsilon$; or equivalently, for any $E \in \mathcal{B}_{\mathbb{R}^n}$,

$$P(E) = \sup\{P(K) : K \subseteq E, K \text{ compact}\}$$

$$= \inf\{P(U) : U \supseteq E, U \text{ open}\}.$$

(*Hint:* Mimic the proof given for the case $n = 1$; just use the fact that $\mathcal{B}_{\mathbb{R}^n} = \sigma(\mathcal{A})$, where a typical set in $\mathcal{A}$ is a disjoint union of sets of the form $\prod_{i=1}^{n} [a_i, b_i)$.)

Our interest in regularity stems from the fact that if $\{K_n\}_{n=1}^{\infty}$ is a sequence of non-empty compact sets such that $K_1 \supseteq K_2 \supseteq \ldots$, then $\cap_n K_n \neq \phi$.

Theorem 4.4.4   (**Kolmogorov consistency theorem**) *Suppose P_F is a probability measure defined on $(\Omega, \mathcal{B}_F)$ for each F in $\mathcal{F}_I$. Suppose the measures $\{P_F : F \in \mathcal{F}_I\}$ are consistent, meaning that $P_G | \mathcal{B}_F = P_F$ whenever $F, G \in \mathcal{F}_I$ and $F \subseteq G$. Then there exists a unique probability measure P defined on $(\Omega, \mathcal{B})$ such that $P | \mathcal{B}_F = P_F$ for all F in $\mathcal{F}_I$.*

Proof: Uniqueness is a consequence of Theorem 1.5.11 and the fact that $\mathcal{A} = \cup_{F \in \mathcal{F}_I} \mathcal{B}_F$ is an algebra such that $\mathcal{B} = \sigma(\mathcal{A})$.

As for existence, define $P : \mathcal{A} \to [0, 1]$ by the prescription $P | \mathcal{B}_F = P_F$. This is well defined by the assumed consistency. Thanks to Caretheodory's extension theorem and Lemma 4.3.4, we only need to verify that if $\{A_n\}_{n=1}^{\infty}$ is a decreasing sequence of sets such that $A_n \in \mathcal{A}$ $\forall n$ and $\cap_n A_n = \phi$, then $\lim_{n \to \infty} P(A_n) = 0$. Before proceeding to do this, we make a few simplifications:

Step 1: We may, without loss of generality, assume that $I = \mathbb{N} = \{1, 2, 3, \ldots\}$.

(Reason: Pick $F_n \in \mathcal{F}_I$ such that $A_n \in \mathcal{B}_{F_n}$. Since each F_n is finite, the union $I_0 = \bigcup_{n=1}^{\infty} F_n$ is countable, and as far as verifying that $\lim_{n \to \infty} P(A_n) = 0$ is concerned, it suffices to treat the case $I = I_0$.)

Thus, we assume that $\Omega = \mathbb{R}^{\mathbb{N}}$; as before, we write $\mathcal{B}_{n]}$ for $\mathcal{B}_{\{1,2,\ldots,n\}}$. Then $\mathcal{A} = \bigcup_{n=1}^{\infty} \mathcal{B}_{n]}$.

Step 2: We may, without loss of generality, assume that $A_n \in \mathcal{B}_{n]}$ for each n.

(Reason: If $A_n \in \mathcal{B}_{m:n]}$, and $k_n = \max\{m_1, \ldots, m_n\}$, define $\tilde{A}_1 = \tilde{A}_2 = \cdots = \tilde{A}_{k_1-1} = \Omega$; $\tilde{A}_{k_1} = \tilde{A}_{k_1+1} = \cdots = \tilde{A}_{k_2-1} = A_1$; $\tilde{A}_{k_2} = \tilde{A}_{k_2+1} = \cdots = \tilde{A}_{k_3-1} = A_2$, etc. Then $\tilde{A}_n \in \mathcal{B}_{n]}$, $\tilde{A}_1 \supseteq \tilde{A}_2 \supseteq \ldots$, $\bigcap_{n=1}^{\infty} \tilde{A}_n = \phi$ and $\lim_{n \to \infty} P(\tilde{A}_n) = \lim_{n \to \infty} P(A_n)$.)

Thus we assume $\Omega = \mathbb{R}^{\mathbb{N}}$ and $A_n = \pi_{n]}^{-1}(E_n)$ for some $E_n \in \mathcal{B}_{\mathbb{R}^n}$. The fact that $A_1 \supseteq A_2 \supseteq \ldots$ implies that $P(A_1) \geq P(A_2) \geq \ldots$, and so $\varepsilon = \lim_{n \to \infty} P(A_n)$ exists.

Our strategy is to prove, using inner regularity of probability measures defined on $(\mathbb{R}^n, \mathcal{B}_{\mathbb{R}^n})$, that if $\varepsilon > 0$, then $\bigcap_{n=1}^{\infty} A_n \neq \phi$.

So, suppose $\varepsilon > 0$. Since, by assumption, $P|\mathcal{B}_{n]}$ is a probability measure, we can, by Proposition 4.4.2, find a compact set $C_n \subseteq E_n$ such that $P(C_n) \geq \varepsilon - \varepsilon/2^{n+1}$. Set $B_n = \pi_{n]}^{-1}(C_n)$. Note that if $\tilde{A}_n = \bigcap_{k=1}^{n} B_k$, then $\tilde{A}_n \in \mathcal{B}_{n]}$ and in fact $\tilde{A}_n = \pi_{n]}^{-1}(K_n)$, where $K_n = \{(x_1, \ldots, x_n) \in C_n : (x_1, \ldots, x_k) \in C_k$ for $1 \leq k \leq n\}$ is a compact subset of $\mathbb{R}^n$. An easy estimation (using the fact that $\varepsilon/4 + \varepsilon/8 + \cdots + \varepsilon/2^{n+1} < \varepsilon/2$) shows that $P(\tilde{A}_n) \geq \varepsilon/2$ for all n, and hence each K_n is a non-empty compact set.

Now observe that $\{\pi_1(\tilde{A}_n)\}_{n=1}^{\infty}$ is a decreasing sequence of non-empty compact subsets of $\mathbb{R}$ and hence $\bigcap_{n=1}^{\infty} \pi_1(\tilde{A}_n) \neq \phi$. Pick $x_1^0 \in \bigcap_{n=1}^{\infty} \pi_1(\tilde{A}_n)$. Next, note that if $\tilde{A}_n^{x_1^0} = \{x \in \tilde{A}_n : \pi_1(x) = x_1^0\}$, then each $\tilde{A}_n^{x_1^0} \neq \phi$ and $\{\pi_2(\tilde{A}_n^{x_1^0})\}_{n=1}^{\infty}$ is a decreasing sequence of non-empty compact subsets of $\mathbb{R}$ and hence $\bigcap_{n=1}^{\infty} \pi_2(\tilde{A}_n^{x_1^0}) \neq \phi$. So there exists $x_2^0 \in \mathbb{R}$ such that $(x_1^0, x_2^0) \in \bigcap_{n=1}^{\infty} \pi_{2]}(\tilde{A}_n)$. By repeating this argument indefinitely, we can find a sequence $(x_1^0, x_2^0, x_3^0, x_4^0, \ldots)$ such that $(x_1^0, \ldots, x_n^0) \in \bigcap_{k=1}^{\infty} \pi_{n]}(\tilde{A}_k)$ for each n.

Since $\tilde{A}_n = \pi_{n]}^{-1}(K_n)$, this shows that $(x_1^0, \ldots, x_n^0) \in K_n$ and hence, $(x_1^0, x_2^0, \ldots) \in \tilde{A}_n$ for all n. Thus $\bigcap_{n=1}^{\infty} A_n \supseteq \bigcap_{n=1}^{\infty} \tilde{A}_n \neq \phi$. This contradiction completes the proof. $\square$

One consequence of Kolmogorov's consistency theorem is the existence of the 'product' of an arbitrary family of probability measures on $\mathbb{R}$: explicitly, if $\{P_i : i \in I\}$ is a family of probability measures defined on $(\mathbb{R}, \mathcal{B}_{\mathbb{R}})$, we may, for every finite subset F of I, construct (see Ex. 4.1.9) the product measure $\prod_{i \in F} P_i$ on $\otimes_{i \in F} \mathcal{B}_{\mathbb{R}}$, define P_F to be the corresponding (see Ex. 1.8.3) probability measure on $(\Omega, \mathcal{B}_F)$ such that $P_F \circ \pi_F^{-1} = \prod_{i \in F} P_i$) and invoke Theorem 4.4.4 to deduce the existence of the infinite product $\prod_{i \in I} P_i$.

Rather than going through the details of the above construction of infinite product measures, we list a series of exercises that should introduce the reader to what may be called (continuous time) **Markov processes** (which include the product measure construction as a special case). The reader should note that the construction in the following exercises constitutes a (correct) generalisation to continuous time of the Markov chains that were briefly discussed in the last section.

EXERCISES

Ex. 4.4.5 Let P_n be a probability measure defined on $(\mathbb{R}^n, \mathcal{B}_{\mathbb{R}^n})$. Suppose $P : \mathbb{R} \times \mathcal{B}_{\mathbb{R}} \to [0, 1]$ is a map such that:

(a) $P(y, \bullet)$ is a probability measure defined on $(\mathbb{R}, \mathcal{B}_\mathbb{R})$ for each $y \in \mathbb{R}$; and
(b) $P(\bullet, E)$ is a $(\mathcal{B}_\mathbb{R}, \mathcal{B}_\mathbb{R})$-measurable function for each $E \in \mathcal{B}_\mathbb{R}$.
       Define $P_{n+1} : \mathcal{B}_{\mathbb{R}^{n+1}} \to [0, 1]$ by

$$P_{n+1}(E) = \int_{\mathbb{R}^n} P(y_n, E^{(y_1, \dots, y_n)}) d P_n(y_1, \dots, y_n),$$

where $E^{(y_1, \dots, y_n)} = \{y \in \mathbb{R} : (y_1, \dots, y_n, y) \in E\}$. Show that P_{n+1} is a probability measure defined on $(\mathbb{R}^{n+1}, \mathcal{B}_{\mathbb{R}^{n+1}})$ such that $P_{n+1}(A \times \mathbb{R}) = P_n(A)$ for all $A \in \mathcal{B}_{\mathbb{R}^n}$. (If $X_j : \mathbb{R}^{n+1} \to \mathbb{R}$ denotes the jth projection for $1 \leq j \leq n + 1$, then the quantity $P(y, E)$ is to be thought of as 'the conditional probability that $X_{n+1} \in E$, given that $X_n = y$'.)

Ex. 4.4.6   Suppose $P(s, x; t, E) \in [0, 1]$ whenever $s, x, t \in \mathbb{R}, 0 \leq s < t$ and $E \in \mathcal{B}_\mathbb{R}$. Suppose this assignment satisfies the following requirements:

(i)   if $s, x, t \in \mathbb{R}$ and $0 \leq s < t$, then $P(s, x; t, \bullet)$ is a probability measure defined on $(\mathbb{R}, \mathcal{B}_\mathbb{R})$; ($P(s, x; t, E)$ can be thought of as the probability of transition from point x at time s into the set E at time t) and
(ii)   if $0 \leq s < t$ and $E \in \mathcal{B}_\mathbb{R}$, then $P(s, \bullet; t, E)$ is a $(\mathcal{B}_\mathbb{R}, \mathcal{B}_\mathbb{R})$-measurable function.

Suppose P_0 is a given probability measure defined on $(\mathbb{R}, \mathcal{B}_\mathbb{R})$.

(a)   Suppose there exists a probability measure P defined on $(\Omega, \mathcal{B}) = (\mathbb{R}^{[0,\infty]}, \otimes_{t \in [0,\infty]} \mathcal{B}_\mathbb{R})$ such that:

(i) $P(\{w \in \Omega : w(0) \in E\}) = P_0(E) \ \forall E \in \mathcal{B}_\mathbb{R}$, i.e. $P \circ \pi_0^{-1} = P_0$; and
(ii)   if we write $P_{t_1, \dots, t_n}(E) = P(\{w \in \Omega : (w(t_1), \dots, w(t_n)) \in E\})$ whenever $0 \leq t_1 < t_2 < \cdots < t_n$ and $E \in \mathcal{B}_{\mathbb{R}^n}$, then we have, as in Ex. 4.4.5,

$$P_{t_1, \dots, t_{n+1}}(E) = \int_{\mathbb{R}^n} P(t_n, y_n; t_{n+1}, E^{(y_1, \dots, y_n)})$$

$$\times d P_{t_1, \dots, t_n}(y_1, \dots, y_n) \tag{4.4.4}$$

whenever $0 \leq t_1 < \cdots < t_{n+1}$ and $E \in \mathcal{B}_{\mathbb{R}^n}$.

Then, the 'system $\{P(s, x; t, E)\}$ of transition probabilities' satisfies the following condition (known as the **Chapman–Kolmogorov equations**) whenever $0 \leq s < u < t$ and $E \in \mathcal{B}_\mathbb{R}$.

$$P(s, x; t, E) = \int P(u, y; t, E) P(s, x; u, dy),$$

where we write $\int f(y) P(s, x; u, dy)$ for the integral of f with respect to the measure $P(s, x; u, \bullet)$.

(b)   Show, conversely, (using Ex. 4.4.5) that if the 'system of transition probabilities' satisfies the Chapman–Kolmogorov equations, then there exists a
unique probability measure P defined on $(\Omega, \mathcal{B}) = (\mathbb{R}^{[0,\infty)}, \otimes_{t\in[0,\infty)} \mathcal{B}_{\mathbb{R}})$
satisfying conditions (i) and (ii) of (a) above.

Ex. 4.4.7   If $\{P_t : t \in [0,\infty)\}$ is any family of probability measures on $(\mathbb{R}, \mathcal{B}_{\mathbb{R}})$
such that $t \mapsto P_t(E)$ is a measurable map for each Borel set E, define $P_0 = P_0$
and $P(s, x; t, E) = P_t(E)$ and deduce from Ex. 4.4.6 that there exists a unique
probability measure P on $(\Omega, \mathcal{B}) = (\mathbb{R}^{[0,\infty)}, \otimes_{t\in[0,\infty)} \mathcal{B}_{\mathbb{R}})$ such that

$$P(\{w \in \Omega : w(t_1) \in A_1, \dots, w(t_n) \in A_n\}) = P_{t_1}(A_1) \dots P_{t_n}(A_n)$$

for arbitrary $0 \le t_1 < \cdots < t_n$ and $A_1, \dots, A_n \in \mathcal{B}_{\mathbb{R}}$.

Ex. 4.4.8   In the notation of Ex. 4.4.6, define $P_0 = \delta_0$. Thus $P_0(E) = 1_E(0)$ ($=$
0 or 1 according to whether $0 \notin E$ or $0 \in E$) and

$$P(s, x; t, E) = \frac{1}{\sqrt{2\pi(t-s)}} \int_E \exp\left(-\frac{(y-x)^2}{2(t-s)}\right) dy$$

whenever $s, x, t \in \mathbb{R}, 0 \le s < t$ and $E \in \mathcal{B}_{\mathbb{R}}$. Show that the above specification
satisfies the Chapman–Kolmogorov equations and hence yields a system of
transition probabilities and consequently a unique probability measure P on
$(\mathbb{R}^{[0,\infty)}, \otimes_{t\in[0,\infty)} \mathcal{B}_{\mathbb{R}})$ as in Ex. 4.4.6. If $\{X_t : t \in [0,\infty)\}$ are the random variables defined on $(\mathbb{R}^{[0,\infty)}, \otimes_{t\in[0,\infty)} \mathcal{B}_{\mathbb{R}}, P)$ by $X_t(w) = w(t)$, then $\{X_t : t \ge 0\}$
is the 'stochastic process' that describes the famous Brownian (1-dimensional)
motion.

5

Characteristics and Convergences

In this chapter we will prove the *weak law of large numbers* (Theorem 5.4.3) and the *strong law of large numbers* (Theorem 5.4.26) that provide the mathematical basis for interpretation of probability as 'relative frequency'.

Once probability spaces are constructed, there are instances when one cannot compute the probability of a certain event explicitly. For example, suppose we toss a biased coin n times. Assume that the probability of getting a head is p. Then it can be shown that the

$$P(\{\text{number of heads in } n \text{ tosses is } k\}) = \binom{n}{k} p^k (1 - p)^{n-k}.$$

One can easily observe that for large n it is difficult to compute this explicitly. The way out of this difficulty is to appeal to the fundamental result in probability known as the *central limit theorem*. The central limit theorem holds in a general framework and is a powerful result. It is presented in Theorem 5.3.2. The proof uses the uniqueness theorem (Theorem 5.1.5) and continuity theorem (Theorem 5.3.1).

In order to prove the above results we shall develop a number of tools and techniques that are used widely in probability. We begin with a discussion of characteristic functions.

5.1 Characteristic Functions

Let X be a random variable on a probability space $(\Omega, \mathcal{B}, P)$. Recall that the distribution function $F_X(x) = P_X((-\infty, x]) = P(X \leq x)$ determines the distribution of the random variable X. Another such 'determining' function is the **characteristic function** of the random variable.

Definition 5.1.1 *Let X be a random variable on a probability space $(\Omega, \mathcal{B}, P)$. The characteristic function of X is the function $\phi_X : \mathbb{R} \to \mathbb{C}$ defined by*

$$\phi_X(t) = E(e^{itX}).$$

Note that as $|e^{it}| < 1$, the characteristic function is well defined. Table B.1 in the Appendix provides the formulae for characteristic functions of the standard distributions discussed in the previous sections. Here we compute the characteristic functions for the geometric and exponential random variables.

Example 5.1.2 *Let X be a geometric (p) random variable. Then, for any $t \in \mathbb{R}$, its characteristic function $\phi(t)$ is given by*

$$\phi(t) = Ee^{itX}$$

$$= \sum_{k=1}^{\infty} p(1-p)^{k-1} e^{itk}$$

$$= \frac{p}{1-p} \sum_{k=1}^{\infty} (1-p)^k e^{itk}$$

$$= \frac{p}{(1-p)} \frac{(1-p)e^{it}}{(1-(1-p)e^{it})}$$

$$= \frac{pe^{it}}{1-(1-p)e^{it}}.$$

Example 5.1.3 *Let X be an exponential (λ) random variable. Then, for any $t \in \mathbb{R}$, its characteristic function $\phi(t)$ is given by*

$$\phi(t) = Ee^{itX}$$

$$= \int_0^\infty dx e^{itx} \lambda e^{-\lambda x}$$

$$= \lambda \left(\int_0^\infty dx \cos(tx) e^{-\lambda x} + i \int_0^\infty dx \sin(tx) e^{-\lambda x} \right)$$

$$= \frac{\lambda}{\lambda^2 + t^2} \left(e^{-\lambda x}(-\lambda \cos(tx) + t \sin(tx)) \Big|_0^\infty \right.$$

$$\left. + e^{-\lambda x}(-t \cos(tx) - \lambda \sin(tx)) \Big|_0^\infty \right)$$

$$= \frac{\lambda}{\lambda - it}.$$

We conclude this section with two important results concerning characteristic functions. First is an inversion formula.

Theorem 5.1.4   (Inversion Theorem) *Let X be a random variable with characteristic function $\phi_X(\cdot)$. Assume that $\int_{\mathbb{R}} |\phi_X(t)| dt < \infty$. Then X has a density function $f : \mathbb{R} \to \mathbb{R}$ given by*

$$f(x) = \frac{1}{2\pi} \int_{\mathbb{R}} e^{itx} \phi_X(t) dt. \tag{5.1.1}$$

The second is a uniqueness theorem.

Theorem 5.1.5   (Uniqueness Theorem) *Two random variables X and Y have the same distribution if and only if $\Phi_X(t) = \Phi_Y(t)$ for all t.*

For proving the above theorems we will need a couple of technical lemmas.

Lemma 5.1.6   *Let X be a real valued random variable with characteristic function $\phi_X(\cdot)$. Let $Z \overset{d}{=} N(0,1)$ be independent of X. For each $\sigma > 0$, the random variable $X_\sigma = X + \sigma Z$ has a density f_σ given by*

$$f_\sigma(x) = \frac{1}{2\pi} \int_{-\infty}^{\infty} e^{itx} \phi_X(t) e^{-\frac{\sigma^2 t^2}{2}} dt \tag{5.1.2}$$

for all $x \in \mathbb{R}$.

Proof: Fix $\sigma > 0$. Using the independence of X and Z, we have

$$P(X_\sigma \le \alpha) = \int_{\mathbb{R}} F_Z(\frac{\alpha - x}{\sigma}) d\mu_X(x)$$

$$= \int_{\mathbb{R}} \int_{-\infty}^{\frac{\alpha-x}{\sigma}} \frac{1}{\sqrt{2\pi}} e^{-\frac{a^2}{2}} da \, d\mu_X(x)$$

$$= \int_{\mathbb{R}} \int_{-\infty}^{\alpha} \frac{1}{\sqrt{2\pi}\sigma} e^{-\frac{(y-x)^2}{2\sigma^2}} dy \, d\mu_X(x)$$

$$= \int_{-\infty}^{\alpha} \int_{\mathbb{R}} \frac{1}{\sqrt{2\pi}\sigma} e^{-\frac{(y-x)^2}{2\sigma^2}} d\mu_X(x) dy. \tag{5.1.3}$$

Let $Y \overset{d}{=} N(0, \frac{1}{\sigma^2})$ be a random variable independent of X and Z. So $\phi_Y(t) = e^{-\frac{t^2}{2\sigma^2}}, t \in \mathbb{R}$. Using the independence of X and Y and the definition of a characteristic function, we have

$$\int_{\mathbb{R}} \frac{1}{\sqrt{2\pi}\sigma} e^{-\frac{(x-a)^2}{2\sigma^2}} d\mu_X(x) = \frac{1}{\sqrt{2\pi}\sigma} \int_{\mathbb{R}} \phi_Y(x - a) d\mu_X(x)$$

$$= \frac{1}{\sqrt{2\pi}\sigma} E(e^{iXY} e^{-iaY})$$

$$= \frac{1}{\sqrt{2\pi}\sigma} E(\phi_X(Y)e^{-iaY})$$

$$= \int_{\mathbb{R}} e^{-iay} \phi_X(y) e^{-\frac{\sigma^2 y^2}{2}} dy. \tag{5.1.4}$$

Now (5.1.4), (5.1.3) and the definition of a density function imply the result.   □

Lemma 5.1.7   *Let μ_1 and μ_2 be two probability measures on $(\Omega, \mathcal{B})$ and $\mathcal{C} = \{f : \mathbb{R} \to \mathbb{R} : f(x) = \frac{1}{\sqrt{2\pi}\sigma} e^{-\frac{1}{2}(\frac{x-a}{\sigma})^2}, a, \sigma \in \mathbb{R}\}$. Suppose*

$$\int f d\mu_1 = \int f d\mu_2, \forall f \in \mathcal{C}. \tag{5.1.5}$$

then $\mu_1 = \mu_2$.

Proof: It is clear that (5.1.5) holds for the vector space $\mathcal{V}$, generated by $\mathcal{C}$ over the real numbers. The Stone–Weierstrass theorem (Theorem A.4.8) tells us that $\mathcal{V}$ is dense in $\mathcal{C}_0(\mathbb{R})$, the class of continuous functions vanishing at ∞. Since an indicator function of an open set can be written as a increasing limit of functions in $\mathcal{C}_0(\mathbb{R})$, the monotone convergence theorem (Theorem 2.2.7) will imply that μ_1 and μ_2 agree on all open sets. As μ_1 and μ_2 are probability measures and the open sets generate the Borel σ-algebra, Proposition 4.4.1 will imply that $\mu_1(B) = \mu_2(B)$ for all Borel sets B.   □

   We are now ready to prove the two theorems.

Proof of Theorem 5.1.4:   Let f be defined by (5.1.1). The hypothesis on $\phi_X(\cdot)$ implies that f is a bounded complex function. Fix $\sigma > 0$. Consider X_σ as in Lemma 5.1.6 and denote its density by $f_\sigma(\cdot)$. Now,

$$|f(x) - f_\sigma(x)| = \frac{1}{2\pi} \left| \int_{\mathbb{R}} e^{itx} \phi_X(t)(1 - e^{-\frac{\sigma^2 t^2}{2}}) \right| dt$$

$$\leq \int_{\mathbb{R}} |\phi_X(t)|(1 - e^{-\frac{\sigma^2 t^2}{2}}) dt.$$

Now for all $t \in \mathbb{R}$, $|\phi_X(t)|(1 - e^{-\frac{\sigma^2 t^2}{2}})$ goes to zero as $\sigma \to 0$ and is less than or equal to $|\phi_X(t)|$. The dominated convergence theorem then shows that

$$\sup_{x \in \mathbb{R}} |f_\sigma(x) - f(x)| \to 0. \tag{5.1.6}$$

So f is real-valued.

Let $a \leq b \in \mathbb{R}$. Define a sequence of functions $g_n : \mathbb{R} \to \mathbb{R}$ by

$$
g_n(x) = \begin{cases}
n(x - a) & \text{if } x \in [a, a + \frac{1}{n}] \\[2mm]
1 & \text{if } x \in [a + \frac{1}{n}, b] \\[2mm]
-n(x - b - \frac{1}{n}) & \text{if } x \in [b, b + \frac{1}{n}] \\[2mm]
0 & \text{otherwise.}
\end{cases}
$$

Now, $g_n \to g$, with $g = 1_{(a,b]}$ and $X_\sigma \to X$ as $\sigma \to 0$. Using (5.1.6) and applying dominated convergence repeatedly, we obtain

$$
P(a < X \leq b) = E(g(X))
$$
$$
= \lim_{n\to\infty} E(g_n(X))
$$
$$
= \lim_{n\to\infty} \lim_{\sigma\to 0} E(g_n(X_\sigma))
$$
$$
= \lim_{n\to\infty} \lim_{\sigma\to 0} \int_{\mathbb{R}} g_n(x) f_\sigma(x) dx
$$
$$
= \lim_{n\to\infty} \int_{\mathbb{R}} g_n(x) f(x) dx
$$
$$
= \int_{\mathbb{R}} g(x) f(x) dx. \tag{5.1.7}
$$

As the above holds for arbitrary $a \leq b, \in \mathbb{R}$ we can conclude that f is the density of X. (The reader who feels uncomfortable about applying dominated convergence theorem to continuously varying family of functions rather than a sequence should either (i) convince oneself that it is permissible in this context or (ii) restrict oneself to a sequence $\sigma_n \to 0$.) □

Proof of Theorem 5.1.5:   From the definition of characteristic function, it is trivial to observe that if X and Y have the same distribution, then their characteristic functions are the same. For the converse we will show that

$$
E(f(X)) = E(f(Y)), \ \forall f \in \mathcal{C}, \tag{5.1.8}
$$

where $\mathcal{C}$ is as in Lemma 5.1.7; this will imply that X and Y have the same distributions. Let μ_X denote the distribution of X and $f \in \mathcal{C}$. Observe that f is the density of a $N(a, \sigma^2)$ random variable. So

$$E(f(X)) = \int f(x) d\mu_X(x)$$

$$= \int \left(\frac{1}{2\pi} \int e^{-iat - \frac{t^2}{2\sigma^2}} e^{ixt} dt \right) d\mu_X(x)$$

$$= \frac{1}{2\pi} \int e^{-iat - \frac{t^2}{2\sigma^2}} \int e^{ixt} d\mu_X(x) \, dt$$

$$= \frac{1}{2\pi} \int e^{-iat - \frac{t^2}{2\sigma^2}} \phi_X(t) dt,$$

where (i) we have used Theorem 5.1.4 and the formula for the characteristic function of an $N(a, \sigma^2)$ distribution from Table B. 1 (see Appendix) in the second step and (ii) we have used Fubini's theorem (Theorem 4.1.5) in the third step. We leave it for the reader to verify that the hypotheses of the theorem are satisfied.

One can now do a similar calculation for $E(f(Y))$ and get a similar expression in terms of $\phi_Y(t)$. The assumption that the characteristic functions agree will yield (5.1.8).   □

EXERCISES

Ex. 5.1.8   Show that the function f given by (5.1.1) is bounded and continuous on $\mathbb{R}$.

Ex. 5.1.9   Let X be a real valued random variable on $(\Omega, \mathcal{F}, P)$ with characteristic function ϕ. Show that

(a) ϕ is a bounded continuous function with $\phi(0) = 1$
(b)   If $E(|X|^m) < \infty$ for some positive integer m, then show that ϕ is m-times differentiable.

Ex. 5.1.10   Verify the formulae for the characteristic functions given in Table B. 1 of the Appendix.

Ex. 5.1.11   Let $n \in \mathbb{N}$ and X be an $\mathbb{R}^n$ valued random variable on $(\Omega, \mathcal{F}, P)$. Define its characteristic function to be

$$\phi_X(a) = E(e^{i<X.a>}),$$

where $a \in \mathbb{R}^n$ and $< X. a > (\omega) = \sum_{j=1}^{n} X_j(\omega) a_j$.

(a)   Generalise Theorem 5.1.5 to such random vectors.
(b)   For any vector $\alpha \in \mathbb{R}^n$ and matrix $B \in M_{n \times n}(\mathbb{R})$, show that $\phi_{\alpha + BX}(a) = e^{i<\alpha.a>} \phi_X(B^T a)$, where B^T is the transpose of the matrix B. (Vectors are thought of as column vectors.)

(c)   Suppose $X = (X_1, X_2, \ldots, X_n)$, where each $\{X_i : 1 \leq i \leq n\}$ is a real valued random variable on $(\Omega, \mathcal{B}, P)$. Then show that $\{X_i : 1 \leq i \leq n\}$ are independent if and only if

$$\phi_X(a_1, a_2, \ldots, a_n) = \prod_{i=1}^{n} \phi_{X_i}(a_i),$$

where $a_i \in \mathbb{R}$, for $1 \leq i \leq n$.

Ex. 5.1.12   Find the characteristic function of the Gamma distribution with parameters (n, λ).

5.2   Modes of Convergence

Let $X, \{X_n : n = 1, 2, \ldots\}$ be a sequence of random variables on $(\Omega, \mathcal{B}, P)$. In statistical applications one is usually concerned with two types of limiting behaviour. Either we try to determine if a sequence of random variables $\{X_n\}$ converges a.e. to a random variable X or if the distribution functions $\{F_{X_n}\}$ converge point-wise to F_X. We shall begin by defining various modes of convergence, and then proceed to examine the relationships between them.

Definition 5.2.1   *A sequence $\{X_n : n = 1, 2, \ldots\}$ of random variables on Ω is said to*

(i)   **converge almost everywhere** *to X, $(X_n \xrightarrow{a.e.} X)$, if there exists a P-null set N such that $\{X_n(\omega)\}$ converges to $X(\omega)$ whenever $\omega \notin N$;*

(ii)   **converge in probability** *to X, $(X_n \xrightarrow{P} X)$, if for every $\epsilon > 0$,*

$$\lim_{n \to \infty} P(\{\omega : |(X_n - X)(\omega)| > \epsilon\}) = 0 ;$$

(iii)   **converge in the rth mean** *to X, $(X_n \xrightarrow{r} X)$, if $E(|X_n - X|^r) \to 0$; and*

(iv)   **converge in distribution** *to X, $(X_n \xrightarrow{d} X)$, if $F_{X_n}(x) \to F_X(x)$ for all continuity points x of F_X. This mode of convergence is also referred to as* **weak convergence.**

We remark that in (v) of the above definition the random variables $X, \{X_n; n \geq 1\}$ need not be defined on the same probability space.

Proposition 5.2.2   *Let X and $\{X_n : n \geq 1\}$ be random variables on $(\Omega, \mathcal{B}, P)$.*

(a) $X_n \xrightarrow{a.e.} X$ *implies that $X_n \xrightarrow{P} X$.*

(b) $X_n \xrightarrow{r} X$ *for some $r \geq 1$, implies that $X_n \xrightarrow{P} X$.*

(c) $X_n \xrightarrow{P} X$ implies that $X_n \xrightarrow{d} X$.

Proof:

(a)   We are given that

$$1 = P(\lim_{n \to \infty} X_n = X) = P(\cap_{\epsilon > 0} \cup_{m=1}^{\infty} \cap_{n=m}^{\infty} A_n^\epsilon), \qquad (5.2.9)$$

where $A_n^\epsilon = \{|X_n - X| < \epsilon\}$. Let $B_m^\epsilon = \cap_{n=m}^{\infty} A_n^\epsilon$. Let $\epsilon_0 > 0$ be given. Since $B_m^{\epsilon_0}$ is an increasing sequence of sets, $P(B_m^{\epsilon_0}) \uparrow P(\cup_{m=1}^{\infty} B_m^{\epsilon_0})$. Hence from (5.2.9), we see that $\forall \delta > 0$, $\exists N$ such that

$$P(B_m^{\epsilon_0}) > 1 - \delta, \; \forall m \geq N.$$

As $B_m^{\epsilon_0} \subset A_m^{\epsilon_0}$, we have shown that $\forall \delta > 0, \; \exists N$

$$P(|X_m - X| < \epsilon_0) > 1 - \delta, \; \forall m \geq N.$$

As ϵ_0 was arbitrary, $X_n \xrightarrow{P} X$.

(b)   This follows from Tchebychev's inequality, since

$$E|X_n - X|^r \geq E|X_n - X|^r 1_{\{|X_n - X| \geq \epsilon\}} \geq \epsilon^r P(|X_n - X| > \epsilon).$$

(c)   Let $\epsilon > 0$. By definition, $F_{X_n}(t) = P(X_n \leq t)$. Hence

$$\begin{aligned} F_{X_n}(t) &= P(X_n \leq t, |X_n - X| > \epsilon) + P(X_n \leq t, |X_n - X| \leq \epsilon) \\ &\leq P(|X_n - X| > \epsilon) + P(X_n \leq t, |X_n - X| \leq \epsilon) \\ &\leq P(|X_n - X| > \epsilon) + P(X \leq t + \epsilon) \\ &= P(|X_n - X| > \epsilon) + F_X(t + \epsilon). \end{aligned}$$

Similarly, one can show that

$$F_X(t - \epsilon) \leq F_{X_n}(t) + P(|X_n - X| \geq \epsilon).$$

Deduce that

$$\liminf_{n \to \infty} F_{X_n}(t) \geq F_X(t - \epsilon) \text{ and } \limsup_{n \to \infty} F_{X_n}(t) \leq F_X(t + \epsilon). \qquad (5.2.10)$$

Allow $\epsilon \to 0$ and conclude that for all continuity points t of F_X we have

$$\lim_{n \to \infty} F_{X_n}(t) = F_X(t). \qquad \square$$

The following examples show that the converses of the above statements may fail to hold.

Example 5.2.3   *Let* $\Omega = [0, 1]$, $\mathcal{B} = \mathcal{B}_{[0,1]}$, $P(dx) = dx$.

(a)   *Let* $X_n = 1_{[\frac{j}{2^k}, \frac{j+1}{2^k}]}$, *if* $n = 2^k + j$, *for some* $j = 0, 1, 2, \ldots, 2^k - 1$
    and $k = 1, 2, \ldots$. *If we let* $A_n = \{X_n > 0\}$, *then clearly,* $P(A_n) \to 0$.
    Consequently $X_n \overset{P}{\to} 0$ *but* $X_n(\omega) \nrightarrow 0$ *for all* $\omega \in \Omega$ *since* $\omega \in A_n$ *i.o.*
(b)   *Let* $X_n = n1_{(0, \frac{1}{n})}$ *and* $X \equiv 0$. *For any* $\epsilon > 0$, *then,*

$$P(|X_n| > \epsilon) \le \frac{1}{n} \, \forall n.$$

Hence $X_n \overset{P}{\to} 0$. *Clearly,* $X_n \nrightarrow X \, \forall r \ge 1$.

Lemma 5.2.4   *If* $X_n \overset{P}{\to} X$, *then* $\exists$ *a subsequence* X_{n_k} *of* X_n *such that* $X_{n_k} \overset{a.e.}{\longrightarrow} X$.

Proof: We are given that $X_n \overset{P}{\to} X$. Choose a subsequence X_{n_k} such that

$$P\left(|X_{n_k} - X| > \frac{1}{k}\right) \le \frac{1}{2^k}.$$

Let $A_k = \{|X_{n_k} - X| > \frac{1}{k}\}$. Clearly, $\sum_{k=1}^{\infty} P(A_k) < \infty$ and so the Borel–Cantelli lemma (Lemma 3.2.2) implies that

$$P(A_k \text{ occur i.o.}) = 0$$

$$\Rightarrow P(\cap_{n=1}^{\infty} \cup_{k=n}^{\infty} A_k) = 0$$

$$\Rightarrow P(\cup_{n=1}^{\infty} \cap_{k=n}^{\infty} A_k') = 1$$

$$\Rightarrow P\left(\left\{\omega \in \Omega : \exists n, \forall k \ge n, |X_{n_k} - X| \le \frac{1}{k}\right\}\right) = 1$$

$$\Rightarrow X_{n_k} \overset{a.e.}{\longrightarrow} X. \qquad \qquad \square$$

Theorem 5.2.5   **(Slutsky's theorem)** *Let* $\{X_n, X, Y_n : n \in \mathbb{N}\}$ *be random variables on a probability space* $(\Omega, \mathcal{B}, P)$. *Let* $X_n \overset{d}{\longrightarrow} X$ *and* $Y_n \overset{P}{\to} c$, *where* $c \in \mathbb{R}$ *(i.e.* $Y_n \overset{P}{\to} Y$ *with* $Y = c$ *a.e.). Then,*

(1) $X_n + Y_n \overset{d}{\longrightarrow} X + c$,

(2) $X_n Y_n \xrightarrow{d} cX,$

(3) $\frac{X_n}{Y_n} \xrightarrow{d} \frac{X}{c}$ *if* $c \neq 0$.

Proof: We shall prove only 1; the other parts can be proved similarly. Let $\epsilon > 0$ be given. Write $F_n = F_{X_n + Y_n}$. Choose t such that $t, t - c + \epsilon, t - c - \epsilon$ are all continuity points of F_X. (This is possible since there can be at most countably many points of discontinuity of F_X.) Now,

$$F_n(t) \leq P(X_n + Y_n \leq t, |Y_n - c| < \epsilon) + P(|Y_n - c| \geq \epsilon)$$

$$\leq P(X_n \leq t - c + \epsilon) + P(|Y_n - c| \geq \epsilon)$$

and

$$F_n(t) \geq P(X_n < t - c - \epsilon) - P(|Y_n - c| \geq \epsilon).$$

So we have shown that

$$F_X(t - c - \epsilon) \leq \liminf_{n \to \infty} F_n(t) \leq \limsup_{n \to \infty} F_n(t) \leq F_X(t - c + \epsilon). \qquad \square$$

We prove the next theorem only for probability measures although it holds for finite measures as well.

Theorem 5.2.6 (Egoroff's theorem) *Let* $\{X_n, n \in \mathbb{N}\}$ *and* X *be random variables on a probability space* $(\Omega, \mathcal{B}, P)$. *If the sequence of random variables* $X_n \xrightarrow{a.e.} X$, *then for every* $\epsilon > 0$ *there exists* $E \in \mathcal{B}$ *such that* $P(E) < \epsilon$ *and* $X_n \to X$ *uniformly on* E'.

Proof: Since all the foregoing notions of convergence are 'translation-invariant', i.e., $X_n \to X \Leftrightarrow X_n - X \to 0$, we may assume without loss of generality that $X \equiv 0$. Suppose $X_n \to 0$ a.e.. Let $F(m, n) = \{\omega : |X_k(\omega)| \geq \frac{1}{m}$ for some $k \geq n\}$ and note that for any m, $\{F(m, n) : n = 1, 2, \ldots\}$ is a decreasing sequence of sets, and

$$P(\cap_{n=1}^{\infty} F(m, n)) = 0.$$

Therefore we may find an integer N_m such that $P(F(m, N_m)) < \frac{\epsilon}{2^m}$. Set $E = \cup_{m=1}^{\infty} F(m, N_m)$ and note that $P(E) < \epsilon$, and that if $\omega \notin E$, then

$$|X_k(\omega)| < \frac{1}{m} \; \forall k \geq N_m.$$

Thus the sequence $\{X_n\}$ converges uniformly on E'. $\qquad \square$

We conclude with a widely applicable result due to Skorokhod. Observe that for a sequence of random variables that converges in distribution, it might not make sense to talk of other modes of convergence. (For instance, these random variables may be defined on different probability spaces.) The theorem however clarifies that there are equivalent copies of the same random variables on a canonical probability space where they converge almost surely.

Theorem 5.2.7   (Skorokhod's theorem) *Let* $X, X_1, X_2, \ldots$ *be a sequence of random variables. The following are equivalent:*

(1) $X_n \xrightarrow{d} X$
(2)   *There exists a probability space* $\{\Omega, \mathcal{B}, P\}$ *and random variables* $Y, Y_1, Y_2, \ldots$ *such that* $Y \stackrel{d}{=} X, Y_n \stackrel{d}{=} X_n$ *for all* $n \in \mathbb{N}$ *and* $Y_n \xrightarrow{a.e.} Y$.

Proof: (2) $\Rightarrow$ (1) is an easy application of Proposition 5.2.2. We shall now assume (1). For $x \in \mathbb{R}$, let us write $F_n = F_{X_n}$ and $F = F_X$. Consider $\Omega = [0, 1]$, $\mathcal{B} = \mathcal{B}_{[0,1]}$, and P= Lebesgue measure on $[0, 1]$. On this probability space define

$$Y_n(\omega) = \inf\{x \in \mathbb{R} : \omega \le F_n(x)\}, \ \forall n \in \mathbb{N}$$

and

$$Y(\omega) = \inf\{x \in \mathbb{R} : \omega \le F(x)\}.$$

Note that $\{\omega \in \Omega : Y(\omega) \le y\} = \{\omega \in \Omega : \omega \le F(y)\}$ and $\{\omega \in \Omega : Y_n(\omega) \le y\} = \{\omega \in \Omega : \omega \le F_n(y)\}$. Thus, $F_Y = F$ and $F_{Y_n} = F_n$, i.e. $Y \stackrel{d}{=} X, Y_n \stackrel{d}{=} X_n$ for all $n \in \mathbb{N}$.

Let $\omega \in \Omega$. Let $\epsilon > 0$ be such that $a = Y(\omega) - \epsilon$ is a continuity point of F. So,

$$Y(\omega) > a \Rightarrow F(a) < \omega$$
$$\Rightarrow \exists m_1 \text{ such that } F_n(a) < \omega \ \forall n \ge m$$
$$\Rightarrow \exists m_1 \text{ such that } Y_n(\omega) > a \ \forall n \ge m,$$

where we have used the fact that $F_n(a) \to F(a)$ in the second step. Therefore,

$$\liminf_{n \to \infty} Y_n(\omega) \ge a = Y(\omega) - \epsilon.$$

The discontinuity points of F being countable, (taking a suitable sequence of $\epsilon_n \to 0$), we have

$$\liminf_{n \to \infty} Y_n(\omega) \ge Y(\omega) \ \forall \omega \in \Omega. \tag{5.2.11}$$

Let $\omega_0 \in \Omega$ be such that $\omega < \omega_0$. Let $\delta > 0$ be such that $b = Y(\omega_0) + \delta$ is a continuity point of F. So,

$$Y(\omega_0) < b \Rightarrow F(b) \geq \omega_0$$

$$\Rightarrow \exists\, m_2 \text{ such that } F_n(b) \geq \omega_0 - \delta \; \forall n \geq m_2$$

$$\Rightarrow \exists\, m_2 \text{ such that } Y_n(\omega_0 - \delta) \leq b \; \forall n \geq m_2,$$

where we have used the fact that $F_n(b) \to F(b)$ in the second step. Therefore

$$\limsup_{n \to \infty} Y_n(\omega_0 - \delta) \leq b = Y(\omega_0) + \delta.$$

As the discontinuity points of F are countable, we first choose $\delta < \omega_0 - \omega$ to obtain (since Y_n is an increasing function)

$$\limsup_{n \to \infty} Y_n(\omega) \leq Y(\omega_0) + \delta.$$

We next choose admissible δ_n such that $\delta_n \to 0$ to obtain

$$\limsup_{n \to \infty} Y_n(\omega) \leq Y(\omega_0). \tag{5.2.12}$$

Note that we have established (5.2.12) for any $\omega < \omega_0$ in Ω. If ω is a continuity point of Y, then allowing ω_0 to decrease to ω yields

$$\limsup_{n \to \infty} Y_n(\omega) \leq Y(\omega). \tag{5.2.13}$$

As the discontinuity points of Y are countable, we conclude from (5.2.11) and (5.2.13) that

$$\lim_{n \to \infty} Y_n = Y(\omega) \;\; a.e.$$

as desired. □

Theorem 5.2.8 (Polya's theorem) *Let F_n be a sequence of distribution functions that converge weakly to a continuous distribution function F. Then,*

$$\lim_{n \to \infty} \sup_{t > 0} |F_n(t) - F(t)| = 0.$$

Proof: Let $\epsilon > 0$ be given. Choose $M > 0$ such that

$$F(M) < \epsilon \text{ and } 1 - F(M) < \epsilon.$$

By our hypothesis, $\exists N_1$ such that

$$|F_n(M) - F(M)| < \epsilon \forall n \geq N_1,$$

$$|F_n(-M) - F(-M)| < \epsilon \forall n \geq N_1. \tag{5.2.14}$$

Hence, for all $t \leq -M$,

$$|F_n(t) - F(t)| \leq F_n(t) + F(t)$$

$$\leq F_n(-M) + F(-M)$$

$$\leq F_n(-M) - F(-M) + 2F(-M).$$

For all $t \geq M$,

$$|F_n(t) - F(t)| = |1 - F(t) - (1 - F_n(t))|$$

$$\leq |1 - F_n(t)| + |1 - F(t)|$$

$$\leq 1 - F_n(M) + 1 - F(M)$$

$$= 1 - F(M) + F(M) - F_n(M) + 1 - F(M).$$

Using (5.2.14) and our choice of M, we have shown that

$$\sup_{|t| \geq M} |F_n(t) - F(t)| < 3\epsilon. \tag{5.2.15}$$

We observe that F is uniformly continuous in the interval $[-M, M]$ and consequently $\exists - M = t_1 \leq t_2 \leq \ldots \leq t_n = M$ such that

$$|F(t_i) - F(t_{i+1})| \leq \epsilon, \ 1 \leq i \leq n - 1.$$

From our hypothesis, let N_2 be such that

$$|F_n(t_i) - F(t_{i+1})| < \epsilon \forall n \geq N_2, 1 \leq i \leq n.$$

If $t \in [-M, M]$ then $t \in [t_i, t_{i+1}]$ for some i.

$$|F_n(t) - F(t)| = \max\{F_n(t) - F(t), F(t) - F_n(t)\}$$

$$\leq \max\{F_n(t_i) - F(t_{i+1}), F(t_{i+1}) - F_n(t_i)\}$$

$$\leq \max\{F_n(t_i) - F(t_i) + F(t_i) - F(t_{i+1}), F(t_{i+1})$$

$$- F(t_i) + F(t_i) - F_n(t_i)\}.$$

Consequently, using (5.2.14), we have that $\forall n \geq N_2$,

$$\sup_{|t| \leq M} |F_n(t) - F(t)| < 2\epsilon. \tag{5.2.16}$$

As $\epsilon > 0$ was arbitrary, (5.2.15) and (5.2.16) imply the result. □

Suppose $\{X_n : n \in \mathbb{N}\}$ and X are continuous random variables with probability density functions f_n and f respectively, and f_n converge pointwise to f. The next result, known as Scheffe's Theorem, implies that the random variables converge in distribution in a uniform sense as in Polya's theorem.

Theorem 5.2.9   **(Scheffe's theorem)** *Let* $f_n(\cdot)$, $f(\cdot)$ *be a sequence of density functions such that* $\lim_{n \to \infty} f_n(t) = f(t) \, \forall t \in \mathbb{R}$. *Then,*

$$\sup_{t \in \mathbb{R}} \left| \int_{-\infty}^{t} (f_n(s) - f(s)) ds \right| \to 0.$$

Proof: Observe that f is integrable, $f_n(s) \wedge f(s) \leq f(s) \forall s \in \mathbb{R}$ and $f_n(s) \wedge f(s) \to f(s) \forall s \in \mathbb{R}$ as $n \to \infty$. So by dominated convergence theorem,

$$\lim_{n \to \infty} \int_{-\infty}^{\infty} f_n(s) \wedge f(s) ds \to \int_{-\infty}^{\infty} f(s) ds = 1.$$

Using this, we have

$$\sup_{t \in \mathbb{R}} \left| \int_{-\infty}^{t} (f_n(s) - f(s)) ds \right| \leq \sup_{t \in \mathbb{R}} \int_{-\infty}^{t} |(f_n(s) - f(s))| ds$$

$$= \int_{-\infty}^{\infty} f_n(s) + f(s) - 2 f_n(s) \wedge f(s) ds$$

$$= 1 + 1 - 2 \int_{-\infty}^{\infty} f_n(s) \wedge f(s) ds$$

$$\to 0 \qquad\qquad \Box$$

EXERCISES

Ex. 5.2.10   Prove (2) and (3) in Theorem 5.2.5.

Ex. 5.2.11   Let $X_n \xrightarrow{d} X$, then show that $X_n^2 \xrightarrow{d} X^2$.

Ex. 5.2.12   Let $Y \overset{d}{=} N(0, 1)$. Let $X_n = (-1)^n Y$. Discuss convergence a.e, in probability, and in distribution of X_n.

Ex. 5.2.13   Let Y_n be a bounded sequence of independent random variables with mean μ and variance σ_n^2. Suppose $Y_n \xrightarrow{a.e.} Y$ and $\sigma_n^2 \to 0$. Then show that $Y = \mu$ a.e.

Ex. 5.2.14   For $n \geq 1$, let $0 \leq p_n \leq 1$. Consider

$$X_n = \begin{cases} 1 & \text{w.p.} \quad p_n \\ 0 & \text{w.p.} \quad 1 - p_n. \end{cases}$$

Let $Y_n = \prod_{k=1}^{n} X_k$. Specify conditions on $\{p_n\}$ that ensure

(a) $Y_n \xrightarrow{p} 0$,

(b) $Y_n \xrightarrow{a.e.} 1$,

(c) $Y_n \xrightarrow{d} Y$, where Y is a random variable

$$Y = \begin{cases} 1 & \text{w.p.} \quad p \\ 0 & \text{w.p.} \quad 1 - p, \end{cases}$$

for some $p \in (0, 1)$.

Ex. 5.2.15   Let $c \in \mathbb{R}$. Let Y_n, Y be a sequence of random variables. Let $f : \mathbb{R} \to \mathbb{R}$ be a continuous function. If $Y_n \xrightarrow{p} c$, then show that $f(Y_n) \xrightarrow{p} f(c)$.

Ex. 5.2.16   Let X_n be a sequence of random variables. Let $a \in \mathbb{R}$. If $X_n \xrightarrow{d} a$ (i.e. $X_n \xrightarrow{d} X$ with $X = a$ a.e.), then show that $X_n \xrightarrow{p} a$.

Ex. 5.2.17   Prove or disprove the following: If $X_n \xrightarrow{d} X$ and $Y_n \xrightarrow{p} Y$, then $X_n + Y_n \xrightarrow{d} X + Y$.

Ex. 5.2.18   Let $Y_1, Y_2, \ldots, Y_n$ be independent random variables, each uniformly distributed over the interval $(0, \theta)$. Show that $\max\{Y_1, \ldots, Y_n\}$ converges in probability to θ as $n \to \infty$.

Ex. 5.2.19   Let $X_n \xrightarrow{d} X$ and let F denote the distribution function of X. Let a be continuity point of F. Show that $P(X_n = a) \to 0$.

Ex. 5.2.20   Let $\{X_n : n \geq 1\}$ be a sequence of random variables that is monotonically increasing, i.e. $X_{n+1}(\omega) \leq X_n(\omega)$ for all $\omega \in \Omega$, $n \in \mathbb{N}$. If $X_n \xrightarrow{p} X$, then show that $X_n \xrightarrow{a.e.} X$.

Ex. 5.2.21   Let Y_n be a sequence of independent and identically distributed (henceforth abbreviated to **i.i.d.**) random variables and let $X_n = \frac{Y_n}{n}$. Show that X_n converges in probability. Decide whether X_n converges a.e. or not.(*Hint*: use Borel–Cantelli lemma.)

Ex. 5.2.22   Let X_n be a sequence of independent random variables on $(\Omega, \mathcal{B}, P)$, such that $X_n \overset{d}{=}$ Exponential (a_n) with $a_n = \ln(n+1)$. Show that the sequence converges to zero in probability. Does the sequence converge to zero almost everywhere ? (*Hint*: use Borel–Cantelli lemma.)

Ex. 5.2.23   Let $\mathcal{F} = \{F : \mathbb{R} \to [0, 1] : F$ is a distribution function.$\}$ Define the function $d : \mathcal{F} \times \mathcal{F} \to [0, \infty)$ by

$$d(F, G) = \inf\{\epsilon > 0 : G(x - \epsilon) - \epsilon \le F(x) \le G(x + \epsilon) + \epsilon\}.$$

Show that $(\mathcal{F}, d)$ is a metric space. Further, show that a sequence of random variables $\{X_n\}$ converges in distribution to X if and only if $\rho(F_{X_n}, F_X) \to 0$.

Ex. 5.2.24   Let $\mathcal{X}$ be the set of all random variables on the probability space $(\Omega, \mathcal{B}, P)$. Define a function $\rho : \mathcal{X} \times \mathcal{X} \to [0, \infty)$ by

$$\rho(X, Y) = E(\min(|X - Y|, 1))$$

for any $X, Y \in \mathcal{X}$. Show that $(\mathcal{X}, \rho)$ is a metric space. Further, show that a sequence of random variables $\{X_n\}$ converges in probability to X if and only if $\rho(X_n, X) \to 0$.

5.3   Central Limit Theorem

In this section we shall prove the central limit theorem. First we prove that the convergence of characteristic functions is equivalent to weak convergence. This is known as the continuity theorem.

Theorem 5.3.1   (Continuity theorem) *Let the random variables $X, \{X_n\}_{n \ge 1}$ have characteristic functions $\phi_X, \{\phi_{X_n}\}_{n \ge 1}$. The following are equivalent:*

1. $X_n \overset{d}{\longrightarrow} X$
2. $E(g(X_n)) \to E(g(X))$ *for all bounded Lipschitz continuous functions,* $g : \mathbb{R} \to \mathbb{R}$. [1]
3. $\lim_{n \to \infty} \phi_{X_n}(t) = \phi_X(t)$, *for all $t \in \mathbb{R}$.*

Proof:

- (1) $\Rightarrow$ (2): is an application of Theorem 5.2.7 and dominated convergence theorem.
- (2) $\Rightarrow$ (3): is obvious by definition of the characteristic function.

---

[1] Recall that this means $\exists$ a constant $C > 0$ such that $|g(x) - g(y)| \le C|x - y|$.

- (3) ⇒(2): Define $Y_n^{(k)} = X_n + \frac{1}{k}Z$ and $Y^{(k)} = X + \frac{1}{k}Z$, where Z is an $N(0, 1)$ random variable which is independent of all X_n and X. Let $\phi_{n,k}(\cdot)$ be the characteristic function of $Y_n^{(k)}$ and ϕ_k be the characteristic function of $Y^{(k)}$. Let $f_{Y_n^{(k)}}$ (resp., $f_{Y^{(k)}}$) be the density of Y_n^k (resp., Y^k) guaranteed by Lemma 5.1.6. Independence of Z and X_n implies that

$$\phi_{n,k}(t) = e^{-\frac{t^2}{2k^2}} \phi_{X_n}(t)$$

and similarly we also have

$$\phi_k(t) = e^{-\frac{t^2}{2k^2}} \phi_X(t).$$

By (5.1.2), we have

$$|f_{Y_n^{(k)}}(a) - f_{Y^{(k)}}(a)| = \left| \int_{-\infty}^{\infty} (\phi_{n,k}(t) - \phi_k(t))e^{iat} dt \right|$$

$$\leq \int_{-\infty}^{\infty} |\phi_{n,k}(t) - \phi_k(t)| dt$$

$$= \int_{-\infty}^{\infty} e^{-\frac{t^2}{2k^2}} |\phi_{X_n}(t) - \phi_X(t)| dt$$

$$\longrightarrow 0$$

by hypothesis and dominated convergence. By Theorem 5.2.9, we conclude that $Y_n^{(k)} \xrightarrow{d} Y^{(k)}$. Let g be a bounded Lipschitz continuous function (with Lipschitz constant C) on $\mathbb{R}$.

$$|E(g(X_n)) - E(g(X))| \leq |E(g(X_n)) - E(g(Y_n^{(k)}))| + |E(g(Y_n^{(k)}))$$

$$- E(g(Y^{(k)}))| + |E(g(Y^{(k)})) - E(g(X))|$$

$$\leq C(E|X_n - Y_n^{(k)}| + E|X - Y^{(k)}|)$$

$$+ |E(g(Y_n^{(k)})) - E(g(Y^{(k)}))|$$

$$= \frac{2CE|Z|}{k} + |E(g(Y_n^{(k)})) - E(g(Y^{(k)}))|.$$

Using $Y_n^{(k)} \xrightarrow{d} Y^{(k)}$ and the fact (1) ⇒ (2), we have that $E(g(X_n)) \to E(g(X))$. Hence (3) ⇒ (2).

- (2) ⇒ (1): Let $x \in \mathbb{R}$ and $k \in \mathbb{N}$. Define

$$f_k(y) = \begin{cases} 1 & y \leq x \\ k(x - y) + 1 & x \leq y \leq x + \frac{1}{k} \\ 0 & y \geq x + \frac{1}{k} \end{cases}$$

and

$$g_k(y) = \begin{cases} 1 & y \leq x - \frac{1}{k} \\ k(x - y) & x - \frac{1}{k} \leq y \leq x \\ 0 & y \geq x. \end{cases}$$

Observe that $g_k \leq 1_{(-\infty,x]} \leq f_k$. Now

$$\limsup_{n \to \infty} F_{X_n}(x) = \limsup_{n \to \infty} E 1_{(-\infty,x]}(X_n)$$

$$\leq \limsup_{n \to \infty} E f_k(X_n)$$

$$= E f_k(X), \tag{5.3.17}$$

since f_k is a bounded Lipschitz function. Similarly,

$$\liminf_{n \to \infty} F_{X_n}(x) = \liminf_{n \to \infty} E 1_{(-\infty,x]}(X_n)$$

$$\geq \liminf_{n \to \infty} E g_k(X_n)$$

$$= E g_k(X). \tag{5.3.18}$$

Observe that $f_k \to 1_{(-\infty,x]}$ and $g_k \to 1_{(-\infty,x)}$ and that they are uniformly bounded by 1. Letting $k \to \infty$ in (5.3.17) and (5.3.18), by dominated convergence theorem, we have $\limsup_{n \to \infty} F_{X_n}(x) \leq P(X \leq x)$ and $\liminf_{n \to \infty} F_{X_n}(x) \geq P(X < x)$. Hence at all continuity points x of F we have shown that $F_{X_n}(x) \to F_X(x)$.   □

We conclude this section with the central limit theorem.

Theorem 5.3.2   (Central limit theorem) *Let $X, X_1, X_2, \ldots$ be i.i.d. random variables. Assume that $\mu = E(X) < \infty$ and $\sigma^2 = E(X - \mu)^2 < \infty$. Define $S_n = \sum_{i=1}^n X_i$. Then,*

$$\frac{S_n - n\mu}{\sqrt{n}\sigma} \xrightarrow{d} Z, \tag{5.3.19}$$

where Z is a standard normal random variable.

Proof: If we let $Y_n = \frac{S_n - n\mu}{\sqrt{n}\sigma}$ and $Z \stackrel{d}{=} N(0, 1)$, then we need, in view of Theorem 5.3.1, to verify that

$$\lim_m \phi_{Y_n}(t) = \phi_Z(t).$$

Now,

$$(\phi_{Y_n}(t)) = E(\exp(it Y_n))) = \left[E\left(\exp\left(i \frac{t}{\sigma\sqrt{n}}(X - \mu)\right)\right)\right]^n$$

while $\phi_Z(t) = e^{-\frac{t^2}{2}}$. We estimate as follows:

We see from Ex. 5.3.3 that

$$\left| E\left(\exp\left(i \frac{t}{\sigma\sqrt{n}}(X - \mu)\right) - 1 - i \frac{t}{\sigma\sqrt{n}}(X - \mu) + \frac{t^2(X - \mu)^2}{2\sigma^2 n}\right)\right|$$

$$\leq E \min \left(\frac{|t(X - \mu)|^3}{6\sigma^3 n^{\frac{3}{2}}}, \frac{t^2(X - \mu)^2}{\sigma^2 n}\right)$$

$$= \frac{t^2}{2\sigma^2 n} g\left(\frac{t}{\sqrt{n}}\right),$$

with $g(a) = E(\min(\frac{|a(X-\mu)^3|}{3\sigma}, 2(X - \mu)^2)$. Let $\epsilon > 0$ be given. Then, the finite variance hypothesis of the theorem and dominated convergence theorem imply the existence of an n_0 large enough so that

$$n \geq n_0 \Rightarrow \left| g\left(\frac{t}{\sqrt{n}}\right)\right| < \sigma^2 \epsilon$$

$$\Rightarrow \left| E\left(\exp\left(i \frac{t}{\sigma\sqrt{n}}(X - \mu)\right) - 1 - i \frac{t}{\sigma\sqrt{n}}(X - \mu)\right.\right.$$

$$\left.\left. + \frac{t^2(X - \mu)^2}{2\sigma^2 n}\right)\right| \leq \frac{t^2}{2n}\epsilon$$

$$\Rightarrow 1 - \frac{t^2}{2n}(1 + \epsilon) \leq E\left(\exp\left(i \frac{t}{\sigma\sqrt{n}}(X - \mu)\right)\right) \leq 1 - \frac{t^2}{2n}(1 - \epsilon)$$

$$\Rightarrow \left(1 - \frac{t^2}{2n}(1 + \epsilon)\right)^n \leq \phi_{Y_n}(t)) \leq \left(1 - \frac{t^2}{2n}(1 - \epsilon)\right)^n$$

$$\Rightarrow e^{-\frac{t^2}{2}(1+\epsilon)} \leq \liminf_{n\to\infty} \phi_{Y_n}(t) \leq \limsup_{n\to\infty} \phi_{Y_n}(t) \leq e^{-\frac{t^2}{2}(1-\epsilon)}.$$

Consequently,

$$\lim_{n\to\infty} \phi_{Y_n}(t) = e^{-\frac{t^2}{2}} = \phi_Z(t).$$ $\qquad\square$

EXERCISES

Ex. 5.3.3 . Let $t \in \mathbb{R}$, $i = \sqrt{-1}$, and $n \in \{0\} \cup \mathbb{N}$. Then show that

$$\left| e^{it} - \sum_{k=0}^{n} \frac{i\, t^k}{k!} \right| \leq \min\left(\frac{|t|^{n+1}}{n+1!}, 2\frac{|t|^n}{n!} \right).$$

Ex. 5.3.4 Let X_n be a sequence of independent and identically $N(\mu, 1)$ distributed random variables. Let $S_n = \sum_{i=1}^{n} X_i$. Define

$$Y_n = \begin{cases} \frac{S_n}{n} & \text{if } |S_n| > n^{\frac{3}{4}} \\ 0 & \text{otherwise.} \end{cases}$$

Show that $\sqrt{n}(Y_n - \mu) \xrightarrow{d} Y^{\mu}$. Find the distribution of Y^{μ}.

Ex. 5.3.5 Let X, X_n be a sequence of random variables on $(\Omega, \mathcal{B}, P)$. Show that $X_n \xrightarrow{d} X$ if and only if $E(\phi(X_n)) \to E(\phi(X))$ for all bounded continuous functions ϕ.

Ex. 5.3.6 Two types of coin are produced at a factory: a fair coin and a biased one that comes up heads 55% of the time. We have one of these coins but do not know whether it is a fair or biased coin. In order to ascertain which type of coin we have, we shall perform the following statistical test. We shall toss the coin 1000 times. If the coin comes up heads 525 or more times, we shall conclude that it is a biased coin. Otherwise, we shall conclude that it is fair. If the coin is actually fair, what is the probability that we shall reach a false conclusion? What would it be if the coin were biased? (*Hint*: Use central limit theorem and the normal tables.)

Ex. 5.3.7 Let $a_n = \sum_{k=0}^{n} \frac{n^k}{k!} e^{-n}$, $n \geq 1$. Using the central limit theorem, evaluate $\lim_{n \to \infty} a_n$.

5.4 Law of Large Numbers

We conclude this chapter by proving the law of large numbers. Formally speaking, the law says that if we have an i.i.d. sequence of random variables $\{X_n\}$, then the averages $\frac{\sum_{i=1}^{n} X_n}{n}$ converge a.e. A remarkable generalisation says that the law of large numbers is true for a stationary sequence. We begin with the definition of a **stationary sequence** and the **invariant σ-algebra** .

Consider a sequence of random variables $\{X_n, n \geq 1\}$ on a probability space $(\Omega, \mathcal{F}, P)$.

Definition 5.4.1 *The sequence X_n is called stationary if*

$$(X_n, X_{n+1}, \dots,) \stackrel{d}{=} (X_1, X_2, \dots) \qquad (5.4.20)$$

for any $n \in \mathbb{N}$.

In other words, a a sequence of random variables X_n is stationary if

$$P(X_1 \leq x_1, \dots, X_m \leq x_m) = P(X_{n+1} \leq x_1, \dots, X_{n+m} \leq x_m),$$
$$(5.4.21)$$

for all $n, m \in \mathbb{N}$, $\{x_i\}_{i=1}^m \in \mathbb{R}$. Another equivalent condition is

$$P((X_1, X_2, \dots) \in B) = P((X_{n+1}, X_{n+2}, \dots) \in B), \qquad (5.4.22)$$

for all $n \in \mathbb{N}$, $B \in \mathcal{B}^{\mathbb{N}}$, where $(\mathbb{R}^{\mathbb{N}}, \mathcal{B}^{\mathbb{N}})$ is the standard countable product space.
It is easy to see that any i.i.d sequence is necessarily a stationary sequence but not conversely.

Definition 5.4.2 *An event $A \in \mathcal{B}$ is said to be invariant if $\exists B \in \mathcal{B}^{\mathbb{N}}$ such that for every $n \geq 1$,*

$$A = \{(X_n, X_{n+1}, \dots,) \in B\}, \qquad (5.4.23)$$

and $\mathcal{I} = \{A \in \mathcal{B} : A$ is invariant $\}$ is called the invariant σ-algebra.

The invariant σ-algebra might seem abstract in definition but there are many events of interest in $\mathcal{I}$. For example, $A = \{\limsup X_n > 0\}$ is an event in the invariant σ-algebra. Another such invariant event of interest in the context of the law of large numbers is $\{\lim_{n\to\infty} \frac{S_n}{n} = a\}$, where $S_n = \sum_{i=1}^n X_i$.
We begin with the weak law of large numbers.

Theorem 5.4.3 **(Weak Law of Large Numbers)** *Let $X, X_1, X_2, \dots$ be a sequence of i.i.d. random variables. Assume that $m = E(X) < \infty$ and $\sigma^2 = E(X - m)^2 < \infty$. Define $S_n = \sum_{i=1}^n X_i$. Then*

$$\frac{S_n}{n} \stackrel{P}{\to} m. \qquad (5.4.24)$$

Proof: Let $\epsilon > 0$ be given. By independence of the X_n and Tchebychev's inequality,

$$P\left(\left|\frac{S_n}{n} - m\right| > \epsilon\right) \leq \frac{E|\frac{S_n}{n} - m|^2}{\epsilon^2} = \frac{\sigma^2}{n\epsilon^2}.$$

As $\epsilon > 0$ was arbitrary, $\frac{S_n}{n} \stackrel{P}{\to} m$. $\qquad\qquad\square$

We now proceed to the generalisation to the stationary case of the strong law of large numbers.

Theorem 5.4.4   *Let X_i be a stationary sequence of random variables such that $E(|X_1|) < \infty$. Define $S_n = \sum_{i=1}^n X_i$. Then,*

$$\frac{S_n}{n} \xrightarrow{a.e.} E(X_1|\mathcal{I}) \tag{5.4.25}$$

and in $\mathcal{L}^1$.

Remark 5.4.5   *If the X_i were an i.i.d sequence, then they are stationary as well. As X_1 is independent of the invariant σ-algebra $\mathcal{I}$, we have $E(X_1|\mathcal{I}) = E(X_1)$ a.e. (by 3.4.3)and hence we have:*

(Strong Law of Large Numbers) *Let $X, X_1, X_2, \ldots$ be a sequence of i.i.d. random variables. Let $m = E(X)$ and assume that $E(|X|) < \infty$ Define $S_n = \sum_{i=1}^n X_i$. Then,*

$$\frac{S_n}{n} \xrightarrow{a.e.} m. \tag{5.4.26}$$

Our proof is based on a simple idea from [18]. Before plunging into the proof of the general case, we first illustrate it in the special case when the sequence X_i is a stationary Bernoulli(p) sequence of random variables. Define $S_n = \sum_{i=1}^n X_i$. Let $\bar{S} = \limsup_{n \to \infty} \frac{S_n}{n}$ and $\underline{S} = \liminf_{n \to \infty} \frac{S_n}{n}$. Clearly, $\bar{S}$ and $\underline{S}$ are non-negative, at most one and consequently integrable random variables. Fix $\epsilon > 0$, then for every k define

$$N_k = \inf\left\{n \in \mathbb{N} : \frac{X_k + X_{k+1} \ldots + X_{k+n-1}}{n} \geq \bar{S} - \epsilon\right\}.$$

The random variable N_k, in some sense, measures how close we are to $\bar{S}$ and our main effort will be to control the size N_k. It is easy to see that N_k is finite a.e. and are all identically distributed (because of stationarity of X_i). Hence we can choose an m such that $P(N_k > m) < \epsilon$ for all k. Define random variables Y_k and N_k^Y by the following mechanism:

$$Y_k = \begin{cases} X_k & \text{if} \quad N_k \leq m \\ 1 & \text{if} \quad N_k > m \end{cases} \tag{5.4.27}$$

$$N_k^Y = \inf\left\{n \in \mathbb{N} : \frac{Y_k + Y_{k+1} \cdots + Y_{k+n-1}}{n} \geq \bar{S} - \epsilon\right\}. \tag{5.4.28}$$

Clearly, $N_k^Y \leq N_k$ and if k is such that $N_k \geq m$, then $N_k^Y = 1$ (since setting $Y_k = 1$ ensures that we are above $\bar{S} - \epsilon$ immediately). So we have

$$N_k^Y \leq m. \text{ a.e.}$$

So for large enough $n \in \mathbb{N}$, we can break up $\sum_{k=1}^n Y_k$ into pieces of lengths at most M such that the average over each piece is at least $\bar{S} - \epsilon$. Then finally stop at the nth term. Then it is clear that

$$\sum_{k=1}^n Y_k \geq (n - m)(\bar{S} - \epsilon). \tag{5.4.29}$$

By our choice of m,

$$E(Y_k) = E(X_k 1(N_k \leq m)) + P(N_k > m) < E(X_k) + \epsilon = E(X) + \epsilon,$$

for any k. Take expectations in (5.4.29) and use the above inequality to obtain

$$n(E(X) + \epsilon) \geq (n - m)(E(\bar{S}) - \epsilon).$$

Divide by n and first let $n \to \infty$ followed by $\epsilon \to 0$, to get

$$E(\bar{S}) \leq E(X). \tag{5.4.30}$$

Let $\widetilde{X_k} = 1 - X_k$. Applying the above argument to $\widetilde{X}$ (verify this), we have

$$E(\widetilde{\bar{S}}) \leq E(\widetilde{X}).$$

Since $\underline{S} = -\widetilde{\bar{S}}$, this implies

$$E(\underline{S}) \geq E(X). \tag{5.4.31}$$

Now, $\underline{S} \leq \bar{S}$ a.e. So only way (5.4.31) and (5.4.30) can hold only if $\underline{S} = \bar{S}$ a.e. Therefore $\lim_{n \to \infty} \frac{S_n}{n}$ exists almost everywhere; let us call it S.

If we assume that the sequence is independent as well, then note that $\text{Var}(\frac{S_n}{n}) = \frac{\sigma^2}{n} \to 0$. But as the sequence is bounded, by dominated convergence, $0 = \lim_{n \to \infty} E((\frac{S_n}{n} - m)^2) = E(S - m)^2$. Proposition 2.3.4 implies that $S = m$ a.e.

Proof of Theorem 5.4.4:   The proof proceeds in three steps.

Step 1: $\lim\limits_{n \to \infty} \dfrac{S_n}{n}$ exists a.e.

Proof: Let $\bar{S} = \limsup\limits_{n \to \infty} \frac{S_n}{n}$. Fix $\alpha, \beta, \epsilon > 0$. Define

$$\bar{S}_{(\alpha)} = \min\{\bar{S}, \alpha\}, \quad X^{(\beta)} = \max\{X, -\beta\} \text{ and } X_k^{(\beta)} = \max\{X_k, -\beta\} \, \forall k \in \mathbb{N}.$$

Note that expectations of all of the above random variable exists. (*Reason:* By Fatou's lemma and stationarity we have,

$$E \liminf_{n \to \infty} \sum_{k=1}^{n} \frac{|X_k|}{n} \leq \liminf_{n \to \infty} \sum_{k=1}^{n} \frac{E|X_k|}{n} = E|X_1|.$$

The inequality $\frac{\sum_{k=1}^{n} -X_k}{n} \leq \frac{\sum_{k=1}^{n} |X_k|}{n}$ $\forall n \in \mathbb{N}$ implies $-\bar{S} \leq \liminf_{n \to \infty} \sum_{k=1}^{n} \frac{|X_k|}{n}$. Consequently, we have that $E(\bar{S})$ exists, $E(\bar{S}) \geq -E|X_1|$. Now $X_k^{(\beta)}$ is integrable since $X_k^{(\beta)} \leq |X_k|$. Finally, $E(\bar{S}_{(\alpha)})$ makes sense since $\bar{S}_{(\alpha)} \leq \alpha$.)

For every $k \geq 1$ define

$$N_k (\equiv N_k(\alpha, \beta)) = \inf \left\{ n \in \mathbb{N} : \frac{X_k^{(\beta)} + X_{k+1}^{(\beta)} \cdots + X_{k+n-1}^{(\beta)}}{n} \geq \bar{S}_{(\alpha)} - \epsilon \right\}.$$

The random variables N_k are finite a.e. Since the N_k's are identically distributed (because of stationarity), we may, and do choose $m \geq 1$ so that

$$P(N_k > m) < \epsilon \quad \text{for all } k.$$

As in the proof for the Bernoulli(p) case, define random variables Y_k and N_k^Y by:

$$Y_k = \begin{cases} X_k^{(\beta)} & \text{if } N_k \leq m \\ \max(\alpha, X_k^{(\beta)}) & \text{if } N_k > m \end{cases} \tag{5.4.32}$$

$$N_k^Y = \inf \left\{ n \in \mathbb{N} : \frac{Y_k + Y_{k+1} \cdots + Y_{k+n-1}}{n} \geq \bar{S}_{(\alpha)} - \epsilon \right\}. \tag{5.4.33}$$

By definition, $Y^k \geq X_k^{(\beta)}$ which implies $N_k^Y \leq N_k$. Secondly, if k is such that $N_k > m$, then

$$Y_k = \max(\alpha, X^{(\beta)}) \geq \alpha \geq \bar{S}_{(\alpha)} \geq \bar{S}_{(\alpha)} - \epsilon.$$

So $N_k^Y = 1$ in this case. Hence we have

$$N_k^Y \leq m. \quad \text{a.e.}$$

So for large enough $n \in \mathbb{N}$ we can break up $\sum_{k=1}^{n} Y_k$ into chains of length at most m such that the average over each piece is at least $\bar{S}_{(\alpha)} - \epsilon$. Then it is clear that

$$\sum_{k=1}^{n} Y_k \geq (n - m)(\bar{S}_{(\alpha)} - \epsilon) + m(-\beta)$$

$$\geq (n - m)(\bar{S}_{(\alpha)}) - (n - m)\epsilon - m\beta. \tag{5.4.34}$$

By our choice of m,

$$E(Y_k) = E(X_k^{(\beta)} 1_{N_k \leq m}) + E(\max(\alpha, X_k^{(\beta)}) 1_{N_k > m})$$

$$= E(X_k^{(\beta)}) + E((\max(\alpha, X_k^{(\beta)}) - X^{(\beta)}) 1_{N_k > m})$$

$$\leq E(X_k^{(\beta)}) + (\alpha + \beta) P(N_k > m)$$

$$\leq E(X_k^{(\beta)}) + (\alpha + \beta)\epsilon = E(X^{(\beta)}) + (\alpha + \beta)\epsilon,$$

for any k. Take expectations in (5.4.34) and use the above inequality to obtain

$$n(E(X^{(\beta)}) + (\alpha + \beta)\epsilon) \geq \sum_{k=1}^{n} E(Y_k)$$

$$\geq (n - m)(E\bar{S}_{(\alpha)}) - (n - m)\epsilon - m\beta.$$

Divide the above equation by n and first let $n \to \infty$ followed by $\epsilon \to 0$ to obtain

$$E(\bar{S}_{(\alpha)}) \leq E(X^{(\beta)}). \tag{5.4.35}$$

Now, $|X^{(\beta)}| \leq |X|$, $E|X| < \infty$ and the dominated convergence theorem implies

$$\lim_{\beta \to \infty} E(X^{(\beta)}) = E(X). \tag{5.4.36}$$

As $\alpha \to \infty$ we have $\bar{S}_{(\alpha)} 1_{\bar{S} \geq 0} \uparrow \bar{S} 1_{\bar{S} \geq 0}$, so monotone convergence and the fact that $-\infty < E\bar{S} 1_{(\bar{S} \leq 0)} \leq 0$ (*reason*: $E(\bar{S}) \geq -E|X_1|$) will imply

$$\lim_{\alpha \to \infty} E\bar{S}_{(\alpha)} = \lim_{\alpha \to \infty} [E(\bar{S}_{(\alpha)} 1_{\bar{S} \leq 0}) + E\bar{S}_{(\alpha)} 1_{\bar{S} \geq 0}]$$

$$= E(\bar{S} 1_{\bar{S} \leq 0}) + \lim_{\alpha \to \infty} E\bar{S}_{(\alpha)} 1_{\bar{S} \geq 0}$$

$$= E(\bar{S} 1_{\bar{S} \leq 0}) + E(\bar{S} 1_{\bar{S} \geq 0})$$

$$= E(\bar{S}). \tag{5.4.37}$$

From (5.4.35), (5.4.36) and (5.4.37) we conclude that

$$E(\bar{S}) \leq E(X).$$ (5.4.38)

Applying the above argument with $-X_k$ in place of X_k, we obtain

$$E(-\underline{S}) \leq E(-X),$$

where $\underline{S} = \liminf_{n\to\infty} \frac{S_n}{n}$, since $\limsup_{n\to\infty} \sum_{k=1}^{n} \frac{-X_k}{n} = -\underline{S}$. So,

$$E(\underline{S}) \geq E(X)$$ (5.4.39)

but by definition, $\underline{S} \leq \bar{S}$ a.e. So by Proposition 2.3.4, (5.4.38) and (5.4.39) can both be true only if $\underline{S} = \bar{S}$ a.e. So the proof of step 1 is complete.

Step 2: Let the limiting random variable in Step 1 be denoted by W. Then

$$\lim_{n\to\infty} E \left| \frac{S_n}{n} - W \right| = 0.$$

Proof: Let $\epsilon > 0$ be given. Fatou's Lemma and stationarity imply

$$E(|W|) \leq \liminf E \left| \frac{S_n}{n} \right| \leq \liminf \sum_{k=1}^{n} \frac{1}{n} E(|X_k|) = E(|X_1|) < \infty.$$

So, there exists N large enough such that

$$\int_{|X_1-W|>N} |X_1 - W| dP < \epsilon.$$ (5.4.40)

By Step 1 and Egoroff's theorem, there exists a set $E \in \mathcal{B}$ such that

$$P(E') < \frac{\epsilon}{N},$$ (5.4.41)

and $\exists N_1 \in \mathbb{N}$ be such that

$$\sup_{\omega \in E} \left| \frac{S_n}{n} - W \right| < \epsilon$$ (5.4.42)

for all $n \geq N_1$. Using stationarity, (5.4.40), (5.4.41) and (5.4.42), we have for all $n \geq N_1$,

$$E\left(\left|\frac{S_n}{n} - W\right|\right) = \int_E \left|\frac{S_n}{n} - W\right| dP + \int_{E'} \left|\frac{S_n}{n} - W\right| dP$$

$$\leq \epsilon P(E) + \int_{E'} \left|\frac{S_n}{n} - W\right| dP$$

$$\leq \epsilon + \int_{E'} \left|\frac{\sum_{k=1}^n (X_k - W)}{n}\right| dP$$

$$\leq \epsilon + \sum_{k=1}^n \frac{1}{n} \int_{E'} |X_k - W| dP$$

$$= \epsilon + \int_{E'} |X_1 - W| dP$$

$$= \epsilon + \int_{E' \cap \{|X_1 - W| \leq N\}} |X_1 - W| dP$$

$$+ \int_{E' \cap \{|X_1 - W| > N\}} |X_1 - W| dP$$

$$\leq \epsilon + N P(E') + \int_{\{|X_1 - W| > N\}} |X_1 - W| dP$$

$$\leq \epsilon + N P(E') + \epsilon$$

$$\leq 3\epsilon.$$

As ϵ was arbitrary, we are done with step 2.

Step 3: $W = E(X_1 | \mathcal{I})$

Proof: As $\{W > w\} \in \mathcal{I}$, this implies that W is $\mathcal{I}$ measurable. By Step 2 and Step 1, for any $A \in \mathcal{I}$ we have

$$\int_A W dP = \lim_{n \to \infty} \int_A \frac{S_n}{n} dP$$

$$= \lim_{n \to \infty} \frac{1}{n} \sum_{k=1}^n \int_A X_k dP$$

$$= \int_A X_1 dP.$$

Therefore $W = E(X_1 | \mathcal{I})$.   □

Remark 5.4.6   *In proving Theorem 5.4.4 we have actually proved an equivalent version of the celebrated result, namely*
(Ergodic theorem) *Let T be a measure-preserving transformation on $(\Omega, \mathcal{B}, P)$ (i.e., $T : \Omega \to \Omega$ is measurable and $P(T^{-1}(A)) = P(A)$ for all $A \in \mathcal{B}$.) Then for any random variable X on $(\Omega, \mathcal{B}, P)$ such that $E|X| < \infty$,*

$$\lim_{n \to \infty} \sum_{k=0}^{n-1} X(T^k(\omega)) = E(X|\mathcal{I}) \ a.e.,$$

where $\mathcal{I} = \{A \in \mathcal{B} : T^{-1}(A) = T(A).\}$

The above result forms the basis of a subject in probability called ergodic theory. For a detailed discussion on this we refer the reader to [5].

EXERCISES

Ex. 5.4.7   Let X_n be a sequence of random variables on $(\Omega, \mathcal{B}, P)$. Show that if X_n are i.i.d, then they are stationary but the converse is not true.

Ex. 5.4.8   Prove that the statement of the ergodic theorem is implied by the Theorem 5.4.4.

Ex. 5.4.9   Let $\mathcal{T}$ be the tail σ-algebra discussed earlier (see Definition 3.2.4). Let $\mathcal{I}$ be the invariant σ-algebra. Show that $\mathcal{T} \subset \mathcal{I}$ and that the inclusion is strict.

6

Markov Chains

In everyday life, we encounter systems which can be in various possible states, e.g., the price of a stock at the Bombay stock exchange or the number of people waiting to pay a telephone bill in Chennai, or the number of tigers alive at Sariska national park. In each of the above cases the system moves from one possible state to another in a random fashion. The above systems have one crucial property which is: given the present, the past is independent of the future.

In this chapter we will try to make this mathematical structure precise.

6.1 Discrete Time MC

Let S denote the set of possible states of a system. We shall assume that S is a countable set. Let $\mathbf{P} = [p_{ij}]_{i,j \in S}$ be a matrix such that $p_{ij} \in [0, 1]$ $\forall i, j \in S$ and $\sum_{k=1}^{\infty} p_{ik} = 1$ $\forall i \in S$. Such a matrix is said to be a stochastic matrix. Let $\mathbb{W} = \mathbb{N} \cup \{0\}$. Let $\mathcal{B}_{\mathbb{W}}(\equiv \mathcal{B}_{S^{\mathbb{W}}})$ and $\mathcal{A}_{\mathbb{W}}$ denote the product σ-algebra on $S^{\mathbb{W}}$ and its generating algebra of sets, respectively, as discussed in Chapter 4. Recall that the typical set $B \in \mathcal{A}_{\mathbb{W}}$ is of the form $B = \prod_{i \in \mathbb{W}} B_i$, where $B_i \subset S$ and $B_i \neq S$ for only finitely many $i \in \mathbb{W}$.

Definition 6.1.1 *Let μ be any probability measure defined on $(S, 2^S)$. Define P to be the probability measure on $(S^{\mathbb{W}}, \mathcal{B}_{\mathbb{W}})$ whose existence and uniqueness are guaranteed by Theorem 4.3.5 in case S is finite, and by Theorem 4.4.4 in the general case, which satisfies*

$$P(B) = \sum_{i_0 \in B_0, i_1 \in B_1, \dots, i_n \in B_n} \mu(i_0) p_{i_0 i_1} p_{i_1 i_2} \cdots p_{i_{n-1} i_n} \tag{6.1.1}$$

whenever $B = \prod_{i \in \mathbb{W}} B_i$, and $n \in \mathbb{W}$ is such that $B_i = S$ for all $i > n$ (we write $\mu(i_0) \equiv \mu(\{i_0\})$).

*For $n \geq 0$, define random variables $X_n : S^{\mathbb{N}} \to S$ by $X_n(\omega) = \omega_n$ for $\omega = (\omega_n)_{n=0}^{\infty}$. Then X_n is said to be a **Markov chain** on state space S with initial distribution μ and **transition matrix P**.*

Remark 6.1.2

(i) *Note that the law of X_n is determined by μ and the matrix $\mathbf{P}$ (see Ex. 6.1.4).*

(ii) *If for $n \in \mathbb{N}, i_0, i_1, \ldots, i_n \in S$ is such that*

$$P(X_{n-1} = i_{n-1}, X_{n-2} = i_{n-2}, \ldots, X_0 = i_0) > 0,$$

then the above definition implies that

$$P(X_n = i_n | X_{n-1} = i_{n-1}, X_{n-2} = i_{n-2}, \ldots, X_0 = i_0)$$
$$= p_{i_{n-1}i_n} = P(X_n = i_n | X_{n-1} = i_{n-1}). \tag{6.1.2}$$

Reason:

$$P(X_n = i_n | X_{n-1} = i_{n-1}, X_{n-2} = i_{n-2}, \ldots, X_0 = i_0)$$

$$= \frac{P(X_n = i_n, X_{n-1} = i_{n-1}, X_{n-2} = i_{n-2}, \ldots, X_0 = i_0)}{P(X_{n-1} = i_{n-1}, X_{n-2} = i_{n-2}, X_0 = i_0)}$$

$$= \frac{\mu(i_0) p_{i_0 i_1} p_{i_1 i_2} \cdots p_{i_{n-1} i_n}}{\mu(i_0) p_{i_0 i_1} p_{i_1 i_2} \cdots p_{i_{n-2} i_{n-1}}}$$

$$= p_{i_{n-1} i_n};$$

and as $P(X_{n-1} = i_{n-1}) > 0$, we also have

$$P(X_n = i_n | X_{n-1} = i_{n-1})$$

$$= \frac{P(X_n = i_n, X_{n-1} = i_{n-1})}{P(X_{n-1} = i_{n-1})}$$

$$= \frac{\sum_{j_0 \in S} \sum_{j_1 \in S} \sum_{j_2 \in S} \cdots \sum_{j_{n-2} \in S} \mu(j_0) p_{j_0 j_1} p_{j_1 j_2} \cdots p_{j_{n-2} i_{n-1}} p_{i_{n-1} i_n}}{\sum_{j_0 \in S} \sum_{j_1 \in S} \sum_{j_2 \in S} \cdots \sum_{j_{n-2} \in S} \mu(j_0) p_{j_0 j_1} p_{j_1 j_2} p_{j_{n-2} i_{n-1}}}$$

$$= \frac{p_{i_{n-1} i_n} \left(\sum_{j_0 \in S} \sum_{j_1 \in S} \sum_{j_2 \in S} \cdots \sum_{j_{n-2} \in S} \mu(j_0) p_{j_0 j_1} p_{j_1 i_2} p_{j_{n-2} i_{n-1}} \right)}{\sum_{j_0 \in S} \sum_{j_1 \in S} \sum_{j_2 \in S} \cdots \sum_{j_{n-2} \in S} \mu(j_0) p_{j_0 j_1} p_{j_1 i_2} p_{j_{n-2} i_{n-1}}}$$

$$= p_{i_{n-1} i_n}.$$

This is what is meant by the statement 'that for a Markov chain, the past and the future are independent given the present' and is referred to as the Markov property of X_n.

(iii)   *The latter part of the proof of the item (ii) above is seen to show that*

$$P(X_n = j | X_{n-1} = i) = p_{ij}$$

if $P(X_{n-1} = i) > 0$; thus the Markov chain defined above is time-homogeneous. Time-inhomogeneous Markov chain can also be defined (as was done for finite state spaces S, in Definition 4.3.6), but we will not discuss them in this chapter.

(iv)   *The entries of **P** are called the (one-step) transition probabilities of the Markov chain X_n and the matrix **P** the (one-step) transition probability matrix of the Markov chain X_n.*

(v)   *Let $\mathbf{P}^n$ be the nth power of the matrix **P** and p_{ij}^n denote the (i, j)th element of $\mathbf{P}^n$. Suppose $i \in S$ is such that $P(X_0 = i) = \mu(i) > 0$.*

a)   *For any $j \in S$, we have*

$$P(X_n = j \mid X_0 = i) = p_{ij}^n \tag{6.1.3}$$

Reason:

$$P(X_n = j | X_0 = i) = \frac{1}{P(X_0 = i)} P(X_n = j, X_0 = i)$$

$$= \frac{1}{\mu(i)} \sum_{j_{n-1} \in S}$$

$$\times P(X_n = j, X_{n-1} = j_{n-1}, X_0 = i)$$

$$= \dots$$

$$= \frac{1}{\mu(i)} \sum_{j_1, \dots, j_{n-1} \in S}$$

$$\times \begin{array}{c} P(X_n = j, X_{n-1} = j_{n-1}, \dots, \\ X_1 = j_1, X_0 = i) \end{array}$$

$$= \frac{1}{\mu(i)} \sum_{j_1, \dots, j_{n-1} \in S}$$

$$\times \mu(i) p_{ij_1} p_{j_1 j_2} \cdots p_{j_{n-2} j_{n-1}} p_{j_{n-1} j}$$

$$= \sum_{j_1, \dots, j_{n-1} \in S} p_{ij_1} p_{j_1 j_2} \cdots p_{j_{n-2} j_{n-1}} p_{j_{n-1} j}$$

$$= p_{ij}^n \quad \text{(by definition).}$$

b)   *For $r, s > 0$ and $r + s = n$, we have*

$$P(X_n = j | X_0 = i) = \sum_{k \in S} P(X_r = k | X_0 = i)$$

$$\times P(X_s = j | X_0 = k). \qquad (6.1.4)$$

This is referred to as the Chapman–Kolmogorov equation.

Reason: This is a consequence of the time-homogeniety implied by part(a), and

$$p_{ij}^n = \sum_{k \in S} p_{ik}^r p_{kj}^s. \qquad (6.1.5)$$

whose verification is left as Ex. 6.1.3.

Notation: For the rest of this chapter we will assume the canonical set up given by Definition 6.1.1. Henceforth, we shall define a Markov chain X_n by specifying its state space S, the distribution μ of X_0, and transition matrix $\mathbf{P}$.

For later reference, if $i \in S$, we shall define P_i to be the measure P corresponding to the initial distribution $\mu = \delta_i$ (the Dirac delta measure at i) and write E_i for expectation with respect to P_i. (Thus, when distributed according to P_i, the Markov chain starts almost surely in the state i.)

EXERCISES

Ex. 6.1.3   Let S be a countable set. Let $P_{|S| \times |S|}$ and $Q_{|S| \times |S|}$ be two stochastic matrices. Then show that the product $R = PQ$ (thus $r_{ij} = \sum_{k \in S} p_{ik} q_{kj}$, $i, j \in S$) makes sense and is also a stochastic matrix.

Ex. 6.1.4   Verify statement (ii) in Remark 6.1.2.

6.2   Examples

We now present some the classical examples of Markov chains.

Example 6.1   **(Gambler's ruin chain)** Suppose Pyare Lal and Lajo have two rupees each. They decide to play a game according to the following rules. At each turn of a coin, if it turns up heads then Pyare Lal will give Lajo one rupee. If it turns up tails then Lajo will give Pyare Lal one rupee. The game ends if any one player runs out of money. Let X_n be the wealth of, say, Lajo at time n. Then X_n is a Markov chain on state space $S = \{0, 1, 2, 3, 4\}$ with $X_0 = 2$ (i.e., $\mu(2) = 1$) and transition matrix

$$\mathbf{P} = \begin{bmatrix} 1 & 0 & 0 & 0 & 0 \\ \frac{1}{2} & 0 & \frac{1}{2} & 0 & 0 \\ 0 & \frac{1}{2} & 0 & \frac{1}{2} & 0 \\ 0 & 0 & \frac{1}{2} & 0 & \frac{1}{2} \\ 0 & 0 & 0 & 0 & 1 \end{bmatrix}.$$

Example 6.2   (Simple symmetric random walk on $\mathbb{Z}$) Consider a frog moving along a river bank according to the following random mechanism. At each time step, it tosses a coin; if heads is the result, it jumps 1 unit up and if tails is the result, it jumps 1 unit down. Let X_n denote the position of the frog at time n. Then X_n is a Markov chain on state space $S = \mathbb{Z}$ with $X_0 = 0$, and transition matrix $\mathbf{P} = [p_{ij}]_{i,j \in \mathbb{Z}}$ given by

$$p_{ij} = \begin{cases} \frac{1}{2} & \text{if } j = i + 1 \\ \frac{1}{2} & \text{if } j = i - 1 \\ 0 & \text{otherwise.} \end{cases}$$

Example 6.3   (Ehrenfest model) Suppose there are totally $2N$ balls distributed equally across two urns. At each time step one of the $2N$ balls is chosen at random and moved from the urn it was in to the other urn. Let X_n be the number of balls in urn 1 at time n. Then X_n is a Markov chain on state space $S = \{0, 1, \dots, 2N\}$ with $X_0 = N$ and transition matrix $\mathbf{P} = [p_{ij}]_{0 \le i,j \le 2N}$ given by

$$p_{ij} = \begin{cases} \frac{i}{2N} & \text{if } j = i - 1, i \ne 0 \\ 1 - \frac{i}{2N} & \text{if } j = i + 1, i \ne 2N \\ 1 & \text{if } (i, j) \in \{(0, 1), (2N, 2N - 1)\} \\ 0 & \text{otherwise.} \end{cases}$$

The above example is referred to as the Ehrenfest model for diffusion of gases.

Example 6.4   (Random walk on graphs) Consider a Markov chain X_n on state space $S = \{1, 2, 3, 4, 5\}$, $X_0 \overset{d}{=}$ uniform $\{1, 2, 3, 4, 5\}$, and transition matrix

$$\mathbf{P} = \begin{bmatrix} 0 & 1 & 0 & 0 & 0 \\ 0.5 & 0 & 0.3 & 0.2 & 0 \\ 0 & 0 & 0 & 0.5 & 0.5 \\ 0 & 0 & 0.5 & 0 & 0.5 \\ 0 & 0 & 0.5 & 0.5 & 0 \end{bmatrix}.$$

X_n models a random walk on S. This chain induces a directed graph with vertex set S. See Figure 6.1 for a graphical representation of this chain.

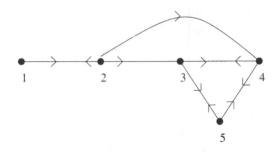

Fig. 6.1   *Graphical representation of the Markov chain in Example 6.4*

Example 6.5   (**Non-Markov**) Let X_n be the simple-symmetric random walk discussed in Example 6.2. If the maximum process M_n is defined by

$$M_n = \max_{0 \leq k \leq n} X_k, \ n \geq 0,$$

then M_n is not a Markov chain as

$$P(M_5 = 1 | M_0 = 0, \ M_1 = 0, \ M_2 = 0, \ M_3 = 1, \ M_4 = 1) \neq P(M_5 = 1 | M_4 = 1),$$

violating (6.1.2).

EXERCISES

Ex. 6.2.1   The examples given below can be modelled by a Markov chain. Determine the state space, initial distribution and the transition matrix for each.

1. Suppose N black balls and N white balls are placed in two urns so that each urn contains N balls. At each step one ball is selected at random from each urn and the two balls interchange places. The state of the system at time $n \in \mathbb{N}$ is the number of white balls in the first urn after the nth interchange.

2. Suppose a gambler starts out with a certain initial capital of N rupees and makes a series of 1 rupee bets against the gambling house until her capital runs out. Assume that she has probability p of winning each bet. Let the state of the system at time $n \in \mathbb{N}$ denote her capital at the nth bet.

3. A particle is moving along the graph shown below. At each time step it moves along one of the incident edges to a neighbouring vertex, choosing the edge with equal probability and independently of all previous movements. Assume that it starts at a uniformly chosen point on the graph. Let the state of the system at time $n \in \mathbb{N}$ be the position of the particle at time n.

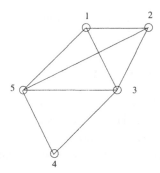

Ex. 6.2.2   Meteorologist Chakrapani could not predict rainy days very well in the wet city of Cherapunjee. So he decided to use the following prediction model for rain. If it had rained yesterday and today, then it will rain tomorrow with probability 0.5. If it rained today but not yesterday, then it will rain tomorrow with probability 0.3. If it did not rain today but had rained yesterday, then it will rain tomorrow with probability 0.1. Finally if it did not rain today and had not rained yesterday, then it will rain tomorrow with probability 0.1. Let X_n denote R if it rained on day $n \in \mathbb{N}$ and D if it was a dry day (no rain). Assume that with probability 0.5 it rains on day 0.

Show that $\{X_n : n \geq 0\}$ is not a Markov chain but $\{Y_n = (X_n, X_{n-1}), n \geq 1\}$ is a Markov chain. Write down the state space, initial distribution and the transition matrix for the chain Y_n.

Ex. 6.2.3   Consider a Markov chain X_n on state space $\{A, B, C\}$ with initial distribution μ and transition matrix given by

$$\mathbf{P} = \begin{pmatrix} .2 & .4 & .4 \\ .4 & .4 & .2 \\ .4 & .6 & 0 \end{pmatrix}.$$

1. What is the probability of going from state A to state B in one step?
2. What is the probability of going from state B to state C in exactly two steps?
3. What is the probability of going from state C to state A in exactly two steps?
4. What is the probability of going from state C to state A in exactly three steps?
5. Calculate the second, third and fourth power of this matrix. Do you have a guess for $\mathbf{P}^n$ for large n (You will find that $\lim_{n \to \infty} p_{ij}^n = \pi(j)$ for $i, j \in S$ for some vector $(\pi(j))_{j \in S}$ which is called the stationary distribution. This will be discussed later in this chapter.)

Ex. 6.2.4 Let X_n and M_n be as in Example 6.5. Verify that the sequence of vectors $Y_n = (X_n, M_n)$ is a Markov chain.

6.3 Classification of States

In Example 6.4, if the chain starts at (or reaches) the states 3, 4, or 5, then it never visits the states 1 or 2. Hence some states in S are not always (i.e. at all times) accessible by the chain with positive probability. In this section we shall try to identify and classify states based on the extent of their accessibility from other states.

Definition 6.3.1   *Let X_n be a Markov Chain on S with transition matrix* **P**.

1. *A state $j \in S$ is said to be* **accessible** *from another state $i \in S$ if $\exists$ a non-negative integer N such that $p_{ij}^N > 0$. This is denoted by $j \leftarrow i$.*
2. *Two states i and j in S are said to* **communicate** *if i and j are both accessible from each other. This is denoted by $j \leftrightarrow i$.*
3. *A subset $C \subset S$ is said to be a* **closed communicating class** *if $i \leftrightarrow j$ and $k \nleftarrow j$ whenever $i, j \in C$ and $k \in S \backslash C$.*
4. *X_n is said to be* **irreducible** *if $i \leftrightarrow j$ for all $i, j \in S$.*
5. *The* **period** *of $i \in S$ is defined as*

$$d(i) = \text{g.c.d.}\{m \geq 1 : p_{ii}^m > 0\} \tag{6.3.6}$$

If $p_{ii}^m = 0$ for all $m \geq 1$, then $d(i)$ is defined to be 0. Finally, X_n is said to be **aperiodic** *if $d(i) = 1$ for all $i \in S$.*

In Example 6.4, $2 \leftrightarrow 1$, $3 \leftarrow 2$, $4 \leftarrow 2$, and $\{3, 4, 5\}$ is a closed communicating class for the Markov Chain. Consequently the chain is not irreducbile.

Lemma 6.3.2 $\leftrightarrow$ *defined above is an equivalence relation on S.*

Proof: This is left as an exercise.   □

Lemma 6.3.3   *Let X_n be a Markov chain on S with transition matrix* **P**. *For $i, j \in S$, if $i \leftrightarrow j$, then $d(i) = d(j)$.*

Proof: As $i \leftrightarrow j$, there exists $r, s \in \mathbb{N}$ such that $p_{ji}^r > 0$ and $p_{ij}^s > 0$. (6.1.5) implies that $p_{jj}^{r+s} > 0$ and $p_{ii}^{s+r} > 0$. Hence $d(j)$ and $d(i)$ are not zero and they both divide $r + s$.

Let $n_0 \in \mathbb{N}$ be such that $p_{ii}^{n_0} > 0$. Again by (6.1.5), we conclude that $p_{jj}^{r+n_0+s} > 0$. Therefore $d(j)$ divides $r + s + n_0$ and by earlier assertion, $d(j)$

will divide n_0. By the definition of $d(i)$ we can conclude that $d(j)$ divides $d(i)$. Similarly, we can see that $d(i)$ divides $d(j)$.   □

In applications the following question is of importance: "Are some states of S visited only finitely many times ?"

In this section we shall identify several criteria for deciding whether a state is visited finitely(or infinitely) many times.

We define the number of visits by time n to the state $i \in S$ by

$$N_n(i) = \sum_{m=0}^{n} 1_{\{X_m=i\}},$$

and the total number of visits to the state $i \in S$ by

$$N(i) = \sum_{n=0}^{\infty} 1_{\{X_n=i\}}.$$

For any $j \in S$, the so-called (first) **hitting time** of the state $j \in S$ is defined by

$$T_j = \inf\{n \geq 1 : X_n = j\}. \tag{6.3.7}$$

If $P \equiv P_j$, then T_j is referred to as the return time to state j.

Definition 6.3.4   *A state $i \in S$ is said to be*

1. **recurrent** *if $P_i(N(i) = \infty) = 1$.*
2. **transient** *if $P_i(N(i) < \infty) = 1$.*

The Markov chain X_n is said to be recurrent (resp. transient) if every state is recurrent (resp. transient).

So, if $i \in S$ is visited infinitely many times, then the state i is recurrent. The following lemma establishes a relationship (intuitively seen easily) between recurrence and hitting times.

Proposition 6.3.5   *Suppose X_n is a Markov chain on S and transition matrix P. A state $i \in S$ is recurrent if and only if $P_i(T_i < \infty) = 1$.*

Proof: We shall derive a relationship between $N(i)$ and T_i using the Markov property of X_n. Let $m \in \mathbb{N}$. First note that

$$\{N(i) \geq m\} = \{N(i) \geq m, T_i < \infty\} = \coprod_{k=1}^{\infty}\{N(i) \geq m, T_i = k\}.$$

So,

$$P_i(N(i) \geq m) = \sum_{k=1}^{\infty} P_i(N(i) \geq m, T_i = k)$$

$$= \sum_{N' \equiv \{k \in \mathbb{N}: P_i(T_i = k) > 0\}} P_i(N(i) \geq m, T_i = k)$$

$$= \sum_{k \in N'} P_i(N(i) \geq m | T_i = k) P_i(T_i = k)$$

$$= \sum_{k \in N'} \frac{P_i(N(i) \geq m | X_k = i, X_{k-1} \neq i, \ldots, X_1 \neq i)}{P_i(T(i) = k)}$$

$$= \sum_{k \in N'} P_i(N(i) \geq m - 1) P_i(T_i = k) \quad (\text{why}?) \qquad (6.3.8)$$

$$= P_i(N(i) \geq m - 1) \sum_{k \in N'} P_i(T_i = k)$$

$$= P_i(N(i) \geq m - 1) P_i(T_i < \infty).$$

Iterating the above reasoning, we find that

$$P_i(N(i) \geq m) = [P_i(T_i < \infty)]^m.$$

Therefore we have

$$P_i(N(i) = \infty) = \lim_{m \to \infty} [P_i(T_i < \infty)]^m. \qquad (6.3.9)$$

It follows that i is recurrent if and only if $P_i(T_i < \infty) = 1$. $\qquad \square$

Remark 6.3.6

1. *The above result can be stated equivalently as*

    $$i \in S \text{ is recurrent if and only if } \sum_{n=1}^{\infty} f_{ii}^n = 1, \qquad (6.3.10)$$

    where $f_{ij}^n = P_j(T_i = n)$ for $i, j \in S$.
2. *If in the above result we assume that the chain is irreducible, then*

    $$i \in S \text{ is recurrent} \iff \sum_{n=1}^{\infty} f_{ij}^n = 1 \iff P_j(T_i < \infty) = 1 \forall j \in S$$

    $$(6.3.11)$$

    (see Ex. 6.3.11).

3. *If i is not recurrent then* $0 \le P_i(T_i < \infty) < 1$ *and (6.3.9) then implies that i has to be transient.*

EXERCISES

Ex. 6.3.7   Let X_n be a Markov chain on state space $S = \{1, 2, 3, 4, 5, 6, 7\}$ with transition matrix

$$
\mathbf{P} = \begin{pmatrix}
0 & \frac{1}{2} & \frac{1}{2} & 0 & 0 & 0 & 0 \\
0 & \frac{1}{2} & 0 & \frac{1}{2} & 0 & 0 & 0 \\
0 & 0 & \frac{1}{2} & 0 & \frac{1}{2} & 0 & 0 \\
0 & \frac{1}{4} & \frac{1}{4} & 0 & \frac{1}{4} & \frac{1}{4} & 0 \\
0 & 0 & \frac{1}{2} & 0 & 0 & 0 & \frac{1}{2} \\
0 & 0 & 0 & 0 & 0 & 0 & 1 \\
0 & 0 & 0 & 0 & \frac{1}{2} & 0 & \frac{1}{2}
\end{pmatrix}
$$

1. Sketch the graph induced by this Markov chain on the vertex set S.
2. Determine the closed communicating classes and their periodicity.

Ex. 6.3.8   (Random walk on a circle) Let S be $\{0, 1, 2, \ldots, L\}$. Consider a particle, starting at a uniformly chosen point in S. At each time step, the particle goes to the right with probability p and left with probability $1-p$ with $0 < p < 1$. If it tries go left from 0 it jumps up to L and similarly if it tries to go right from L it jumps to 0.

1. Write down the transition probability matrix when $L = 6$.
2. Sketch the graph induced by this Markov chain on the vertex set S.
3. For general L, show that the chain is irreducible and determine its periodicity.

Ex. 6.3.9   Justify (6.3.8).

Ex. 6.3.10   Let X_n be a Markov chain on S with transition matrix $\mathbf{P}$ and $m \ge 1$, show that

$$
P_i(N(j) \ge m) = P_i(T_j < \infty)(P_j(T_j < \infty))^{m-1}.
$$

(*Hint:* Mimic the proof of (6.3.9).)

Ex. 6.3.11   Let X_n be an irreducible Markov chain on a state space S with transition matrix $\mathbf{P}$. Then show that

$$
P_i(T_i < \infty) = 1 \iff i \in S \text{ is recurrent}
$$

$$
\iff P_j(T_i < \infty) = 1 \; \forall j \in S.
$$

Ex. 6.3.12   Let X_n be a Markov chain on S with transition matrix **P**. Assume that the chain is irreducible and $i \in S$ is reccurent. Then show that that $T_i = \min\{n \geq 1 : X_n = i\}$ a.e.

Ex. 6.3.13   Let X_n be a Markov chain on state space S with transition matrix **P**. Let $i \in S$ have period 1. Then show that there is an n_0 such that $p_{ii}^n > 0$ for all $n \geq n_0$.

6.3.1   Criteria for recurrence and transience

It is often difficult to verify the condition given in (6.3.10). The following results provide other tests for a state to be recurrent.

Proposition 6.3.14   *Let X_n be a Markov chain on S with transition matrix* **P**. *Then $i \in S$ is recurrent if and only if*

$$\sum_{n=0}^{\infty} p_{ii}^n = \infty.$$

Proof: Using (6.1.3), we have

$$p_{ii}^n = P_i(X_n = i) = P_i(X_n = i, T_i \leq n)$$

$$= \sum_{k=1}^{n} P_i(X_n = i, T_i = k)$$

$$= \sum_{N'=\{k\in\mathbb{N}:k\leq n, P_i(T_i=k)>0\}} P_i(X_n = i, T_i = k)$$

$$= \sum_{k\in N'} P_i(X_n = i | T_i = k) P_i(T_i = k)$$

$$= \sum_{k=1}^{n} P_i(X_n = i | T_i = k) P_i(T_i = k)$$

$$= \sum_{k=1}^{n} P_i(X_{n-k} = i) P_i(T_i = k) \quad \text{(why ?)} \tag{6.3.12}$$

$$= \sum_{k=1}^{n} p_{ii}^{n-k} f_{ii}^k. \tag{6.3.13}$$

We shall now use a standard Markov chain technique to complete the proof. First we define for any $s \in (0, 1)$ and $i \in S$, $A_i(s) = \sum_{n=0}^{\infty} p_{ii}^n s^n$ and

$B_i(s) = \sum_{n=1}^{\infty} f_{ii}^n s^n$. Let $C_i(s) = A_i(s)B_i(s)$ for any $s \in (0, 1)$ be the so-called Cauchy product of the two series. We have

$$C_i(s) = \sum_{k=0}^{\infty} \sum_{l=1}^{\infty} p_{ii}^k f_{ii}^l s^{k+l}$$

$$= \sum_{m=1}^{\infty} \left(\sum_{l=1}^{m} p_{ii}^{m-l} f_{ii}^l \right) s^m$$

$$= \sum_{m=1}^{\infty} p_{ii}^m s^m \quad (\text{by } 6.3.13)$$

$$= A_i(s) - 1.$$

Therefore for any $s \in (0, 1)$ we have shown that

$$A_i(s) = \frac{1}{1 - B_i(s)}. \tag{6.3.14}$$

Using (6.3.14) and two applications of monotone convergence theorem, we find that

$$\sum_{n=1}^{\infty} p_{ii}^n = \lim_{s\uparrow 1} A_i(s) = \lim_{s\uparrow 1} \frac{1}{1 - B_i(s)} = \frac{1}{1 - \sum_{n=1}^{\infty} f_{ii}^n}.$$

The result follows from (6.3.10).   □

The following lemma, which is implied by the above proposition, provides an elementary test for transience and a criterion for recurrence.

Lemma 6.3.15 *Let X_n be a Markov chain on S with transition matrix* **P**. *Let $j \in S$.*

1. *If there is a state $k \in S$ such that $k \leftarrow j$ but $j \nleftarrow k$, then j is transient.*

2. *If $i \leftrightarrow j$ and $i \in S$ is recurrent, then j is recurrent.*

Proof:

(1)   Let $m \in \mathbb{N}$ be smallest number such that $p_{jk}^m > 0$ (such an m exists as $k \leftarrow j$). This implies that there exist $i_1, i_2, i_3, \ldots i_{m-1} \in S$ such that $i_l \neq j$ for $l = 1, 2, \ldots m - 1$ and $p_{ji_1} p_{i_1 i_2} \ldots p_{i_{m-1}k} > 0$. Note then that

$$0 < p_{ji_1} p_{i_1 i_2} \cdots p_{i_{m-1} k}$$

$$= P_j(X_1 = i_1, \ldots, X_{m-1} = i_{m-1}, X_m = k)$$

$$\leq P_j(X_1 = i_1, \ldots, X_{m-1} = i_{m-1}, X_m = k, X_n \neq j \forall n > m)$$

$$(\text{since} \quad j \neq k)$$

$$\leq P_j(X_n \neq j \forall n > 0)$$

$$= P_j(T_j = \infty).$$

So by Proposition 6.3.5, we conclude that j is not recurrent and hence transient, by Remark 6.3.6 (3).

(2)   Since $i \leftrightarrow j$ there exists $r, s \in \mathbb{N}$ such that $p_{ji}^r > 0$ and $p_{ij}^s > 0$. So,

$$\sum_{n=1}^{\infty} p_{jj}^n \geq \sum_{k=1}^{\infty} p_{jj}^{r+s+k}$$

$$\geq \sum_{k=1}^{\infty} p_{ji}^r p_{ii}^k p_{ij}^s$$

$$= p_{ji}^r p_{ij}^s \sum_{k=1}^{\infty} p_{ii}^k = \infty,$$

as i is recurrent (by Proposition 6.3.14) and $p_{ji}^r p_{ij}^s > 0$. So, again by Proposition 6.3.14, we may conclude that the state j is recurrent. $\square$

We digress briefly with a discussion of 'two Markov chains'.

Remark 6.3.16   *Suppose X_n and Y_n are Markov chains on state spaces S_X and S_Y with initial distributions μ_X and μ_Y and probability transition matrices $\mathbf{P}_X$ and $\mathbf{P}_Y$ respectively. Thus, according to our prescription, these chains are, respectively, defined on the probability spaces $(S_X^{\mathbb{W}}, \mathcal{B}_{S_X^{\mathbb{W}}}, P_X)$ and $(S_Y^{\mathbb{W}}, \mathcal{B}_{S_Y^{\mathbb{W}}}, P_Y)$. (Recall the notation $\mathbb{W} = \{0\} \cup \mathbb{N}$ introduced earlier in this chapter.) The point of this remark is that there is no need to be so pedantic, and also to observe that we may think of the two Markov chains as being defined on the same probability space. Indeed, let $(\Omega_{(X,Y)}, \mathcal{B}_{(X,Y)}, P_{(X,Y)})$ be the product space $(S_X^{\mathbb{W}} \times S_Y^{\mathbb{W}}, \mathcal{B}_{S_X^{\mathbb{W}}} \otimes \mathcal{B}_{S_Y^{\mathbb{W}}}, P_X \times P_Y)$, and consider the random variables defined on this product space by $(\widetilde{X}_n(\omega_1, \omega_2) = X_n(\omega_1)$ and $\widetilde{Y}_n(\omega_1, \omega_2) = Y_n(\omega_2)$. It is straightforward to verify that*

- *$\widetilde{X}_n$ is a Markov chain on the state space S with the same initial distribution and transition probability matrix as X_n;*

- $\tilde{Y}_n$ *is a Markov chain on the state space S with the same initial distribution and transition probability matrix as Y_n; and further,*
- *the random variables $\{\tilde{X}_m : m \geq 0\}$ and $\{\tilde{Y}_n : n \geq 0\}$ are independent of one another.*

In case $S_X = S_Y = S$, $\mu_X = \delta_i = \mu_Y$ for some $i \in S$, it will be natural to abuse notation and write P_i for each of the resulting measures P_X, P_Y, $P_{(X,Y)}$.

The next result is a very useful technique for establishing recurrence for Markov chains.

Theorem 6.3.17   (Lyapunov function) *Let X_n be an irreducible Markov chain on $S = \{0\} \cup \mathbb{N}$, with transition matrix $\mathbf{P}$. Then,*

(1)   *the chain is transient if and only if there is a bounded non-constant function $f : S \to \mathbb{R}$ such that*

$$\sum_{j=0}^{\infty} p_{ij} f(j) = f(i), \ \forall i \neq 0, \tag{6.3.15}$$

(2)   *the chain is recurrent if there is a function $f : S \to \mathbb{R}$ such that*

$$\lim_{i \to \infty} f(i) = \infty \tag{6.3.16}$$

and

$$\sum_{j=0}^{\infty} p_{ij} f(j) \leq f(i), \ \forall i \neq 0. \tag{6.3.17}$$

Proof: Consider another Markov chain (as in Remark 6.3.16) Y_n on S with transition matrix Q given by

$$q_{ij} = \begin{cases} 1 & \text{if } i = j = 0 \\ 0 & \text{if } i = 0, j \neq 0 \\ p_{ij} & \text{otherwise.} \end{cases}$$

We proceed to prove the theorem assuming the validity of (6.3.18) and (6.3.19) below (about the chain Y_n), whose justification we postpone to the end of the proof of the theorem.

$$\lim_{n \to \infty} q_{i0}^n = P_i(T_0 < \infty), \tag{6.3.18}$$

$$\lim_{n \to \infty} q_{ij}^n = 0 \ \forall j \neq 0. \tag{6.3.19}$$

Let us write T_i^Y for the hitting time of i for the chain Y (just as T_i was the hitting time of i for the chain X). Fix $i \neq 0$, $j \neq 0$.

(a) Let us assume that the chain X_n is transient. Define $f : \mathbb{N} \to \mathbb{R}$ by

$$f(i) = P_i(T_0^Y < \infty).$$

The function f is clearly bounded. The assumed transience of the state 0 for the chain X_n, and (6.3.11) which is consequently applicable, imply that there exists $i_0 \neq 0$ such that $P_{i_0}(T_0 < \infty) < 1$. By definition of Q,

$$P_{i_0}(T_0^Y < \infty) = P_{i_0}(T_0 < \infty) \quad \text{(why?)} \tag{6.3.20}$$

So $f(i_0) < 1$. As $f(0) = 1$, we see that f is non-constant.
Using the fact that $f(0) = 1$ and the Markov property of Y_n, we see that for all $i \neq 0$,

$$f(i) = \sum_{j=0}^{\infty} q_{ij} f(j). \quad \text{(why ?)} \tag{6.3.21}$$

Hence f satisfies (6.3.15), by definition of Q.

For the converse, let us suppose, if possible, that there is a bounded non-constant solution f of (6.3.15) while the chain (X_n, P) is recurrent. In view of Proposition 6.3.5 and (6.3.18), this implies that

$$\lim_{n \to \infty} q_{i0}^n = 1 \tag{6.3.22}$$

For $i \neq 0$,

$$f(i) = \sum_{j=0}^{\infty} q_{ij} f(j).$$

For $i = 0$ the above equality trivially holds. So iterating n times and using (6.1.5), we will have

$$f(i) = \sum_{j=0}^{\infty} q_{ij}^n f(j), \; \forall i \in S, n \in \mathbb{N}. \tag{6.3.23}$$

Fix $i \in S$. We can find random variables Z_n on some probability space with $P(Z_n = j) = q_{ij}^n$ for all $j \in S, n \in \mathbb{N}$. Let $Z \equiv 0$ be the trivial random variable. Then observe that for all $x \in \mathbb{R}$ we have

$$P(Z_n \leq x) = \sum_{S \ni j \leq x} q_{ij}^n$$

$$\to 1_{[0,\infty)}(x)$$

$$= P(Z \leq x);$$

where we have used (6.3.22), (6.3.19) and the finiteness of the set $\{S \ni j \leq x\}$ in the second step.

Thus $Z_n \xrightarrow{d} Z$. So by the implication (1)$\Rightarrow$(2) of Theorem 5.3.1 we conclude from (6.3.23) that

$$f(i) = \lim_{n \to \infty} \sum_{j=0}^{\infty} q_{ij}^n f(j)$$

$$= \lim_{n \to \infty} Ef(Z_n)$$

$$= Ef(Z) = f(0).$$

So f is a constant function which is a contradiction. Hence (X_n, P) is transient.

(b)   Suppose there is an $f : S \to \mathbb{R}$ satisfying (6.3.16) and (6.3.17). Let $\epsilon > 0$ be given. As $\lim_{i \to \infty} f(i) = \infty$, there exists an m such

$$f(i) > \frac{1}{\epsilon} \quad \forall i \geq m. \tag{6.3.24}$$

Pick a constant C sufficiently large so that $C + f(i) > 0$ for all $i < m$. Notice that $f_1 = f + C$ is strictly positive and has the same properties as f. So we may assume that f is strictly positive.
    The inequality

$$\sum_{j=0}^{\infty} q_{ij} f(j) \leq f(i),$$

is valid for $i \neq 0$ by definition of Q and assumption on f and trivially valid for $i = 0$. Iterating this inequality n times and using (6.1.5), we obtain

$$\sum_{j=0}^{\infty} q_{ij}^n f(j) \leq f(i), \qquad \forall i \in S, n \in \mathbb{N}. \tag{6.3.25}$$

Using (6.3.24) and (6.3.25) we have for all $i \in S, n \in \mathbb{N}$,

$$f(i) \geq \sum_{j=0}^{m-1} q_{ij}^n f(j) + \frac{1}{\epsilon} \left(\sum_{j=m}^{\infty} q_{ij}^n \right) = \sum_{j=0}^{m-1} q_{ij}^n f(j) + \frac{1}{\epsilon} \left(1 - \sum_{j=0}^{m-1} q_{ij}^n \right).$$

Now let $n \to \infty$ in the above. From (6.3.18) and (6.3.19), we have

$$f(i) \geq P_i(T_0 < \infty) f(0) + \frac{1}{\epsilon} (1 - P_i(T_0 < \infty)).$$

Rearranging the above we obtain

$$P_i(T_0 < \infty) \geq 1 - \epsilon[f(i) - P_i(T_0 < \infty)f(0)].$$

Let $\epsilon \to 0$; use Ex. 6.3.11 to conclude that the chain (X_n, P) is recurrent.

Justification of (6.3.18): Using (6.3.29) we have

$$\lim_{n \to \infty} q_{i0}^n = \lim_{n \to \infty} \sum_{k=1}^n P_i(T_0^Y = k) q_{00}^{n-k}$$

$$= \lim_{n \to \infty} \sum_{k=1}^n P_i(T_0^Y = k)$$

$$= \lim_{n \to \infty} \sum_{k=1}^n P_i(T_0 = k) \quad \text{(why?)} \tag{6.3.26}$$

$$= P_i(T_0 < \infty).$$

Justification of (6.3.19): Since P is irreducible there is an s such that $p_{j0}^s > 0$ and so by definition of Q, $q_{j0}^s > 0$. This and $q_{0j} = 0$ imply that

$$P_j\left(T_j^Y < \infty\right) \leq 1 - q_{j0}^s < 1.$$

So, from (6.3.9), we have

$$P_j\left(N^Y(j) = \infty\right) = 0 \tag{6.3.27}$$

So,

$$\sum_{n=1}^\infty q_{ij}^n = \sum_{n=1}^\infty P_i(Y_n = j)$$

$$= \sum_{m=0}^\infty P_i(N^Y(j) \geq m) \qquad \text{(by Ex. 6.3.23)}$$

$$= \sum_{m=0}^\infty P_i(T_j^Y < \infty)(P_j(T_j^Y < \infty))^{m-1}$$

$$\qquad\qquad \text{(by Ex. 6.3.10)}$$

$$= P_i(T_j^Y < \infty) \sum_{m=0}^\infty (P_j(T_j^Y < \infty))^{m-1}$$

$$\leq \sum_{m=0}^\infty (1 - q_{j0}^s)^{m-1} < \infty.$$

This in particular implies (6.3.19).   □

Example 6.6   (**Queuing chain**) Consider a railway ticket reservation counter in Chennai. Assume that people arrive at the counter in some random manner. Suppose one person is served in a unit of time. Administrators are usually interested in the number of people waiting in the queue at each time $n = 1, 2, \ldots$. We will now describe a simple Markov chain that models this problem. Let $\xi, \xi_1, \xi_2, \ldots$ be i.i.d random variables such that

$$P(\xi = k) = p_k, \quad k = 0, 1, 2, \ldots, \tag{6.3.28}$$

with $\sum_{k=0}^{\infty} p_k = 1$ (think of ξ_i as the number of people arriving in time unit i). Let X_0 be the number of people in the queue at time 0. Then the number of people in the queue at time $n \geq 1$ can be described by $X_n = \max\{X_{n-1} - 1, 0\} + \xi_n$. It can be verified that X_n is a Markov chain on state space $S = \{0, 1, 2, \ldots\}$ with some initial distribution μ and transition matrix $\mathbf{P}$ given by

$$\begin{pmatrix} p_0 & p_1 & p_2 & p_3 & p_4 & p_5 & \cdots \\ p_0 & p_1 & p_2 & p_3 & p_4 & p_5 & \cdots \\ 0 & p_0 & p_1 & p_2 & p_3 & p_4 & \cdots \\ 0 & 0 & p_0 & p_1 & p_2 & p_3 & \cdots \\ 0 & 0 & 0 & p_0 & p_1 & p_2 & \cdots \\ \vdots & & & \cdots & & \cdots & \cdots \end{pmatrix}.$$

It is not hard to see that the chain is irreducible if

$$0 < p_0 < 1 \text{ and } \exists k > 0 \text{ such that } p_k > 0.$$

Assuming this, we shall try to decide whether the chain is recurrent or transient by using Theorem 6.3.17. As indicated in the theorem, we will try to find Lyapunov functions f satisfying (6.3.15) or (6.3.16) and (6.3.17).

Transience: Define $f : S \to \mathbb{R}$ by $f(i) = a^i$ for some real number $0 < a < 1$. Therefore, we need to find $0 < a < 1$ such that for $i \neq 0$,

$$\sum_{j=0}^{\infty} p_{ij} f(j) = \sum_{j=i-1}^{\infty} p_{j-i+1} a^j = a^i.$$

In other words, dividing both sides of above by a^{i-1}, we need to find a such that

$$a = \sum_{j=i-1}^{\infty} p_{j-i+1} a^{j-i+1}, \ i \neq 0.$$

Let $g(a) = \sum_{k=0}^{\infty} p_k a^k$. Now $g(0) = p_0 > 0$, $g(1) = 1$ and $g'(1) = \sum_{k=0}^{\infty} k p_k = \mu$ (say). Notice that if $\mu > 1$ then there must exist $0 < a_0 < 1$ such that $g(a_0) = a_0$. (*Reason:* The function $g(x) - x$ is positive at $x = 0$, vanishes at $x = 1$ and is strictly increasing at $x = 1$ and therefore negative just to the left of $x = 1$.) So, the function $f(i) = a_0^i$ satisfies (6.3.15), thereby establishing transience of the chain if $\mu > 1$.

Recurrence: Define $f : S \to \mathbb{R}$ by $f(i) = i$. Clearly, f satisfies (6.3.16).

$$\sum_{j=0}^{\infty} p_{ij} f(j) = \sum_{j=i-1}^{\infty} p_{j-i+1} j$$

$$= \sum_{k=0}^{\infty} p_k (k + i - 1)$$

$$= \mu + i - 1$$

So f will satisfy (6.3.17) if $\mu \le 1$ and the chain will be recurrent.

In conclusion, the chain will be transient if $\mu > 1$ and recurrent otherwise.

EXERCISES

Ex. 6.3.18   Prove the assertion that the queuing chain is irreducible if $p_0 > 0$ and if there exists a $k > 0$ such that $p_k > 0$.

Ex. 6.3.19   (Random walk on $\mathbb{Z}$) Let X_n be a Markov chain on $S = \mathbb{Z}$ with transition matrix $\mathbf{P}$ given by

$$p_{ij} = \begin{cases} p & \text{if } j = i + 1 \\ 1 - p & \text{if } j = i - 1 \\ 0 & \text{otherwise,} \end{cases}$$

where $0 < p < 1$.

1.   Show that the chain is irreducible and every state has period 2.
2.   Show that for every $i \in \mathbb{Z}$ $n \in \mathbb{N}$, $p_{ii}^{2n} = \binom{2n}{n} p^n (1 - p)^n$.
3.   Decide whether the chain is recurrent or transient.
4.   What can you say if $p \in \{0, 1\}$?

Ex. 6.3.20   (Simple symmetric random walk on $\mathbb{Z}^d$) Let $S = \mathbb{Z}^d$, and call i a neighbour of j in $\mathbb{Z}^d$, denoted by $i \sim j$ if the Euclidean distance from i to j is one. Let X_n be a Markov chain on S with transition matrix $\mathbf{P}$ given by

$$p_{ij} = \begin{cases} \frac{1}{2d} & \text{if } i \sim j \\ 0 & \text{otherwise.} \end{cases}$$

1. Show that the chain is irreducible and determine its periodicity.
2. Decide whether the chain is recurrent or transient for the cases $d = 2, 3$.
3. What can you say for $d \geq 4$?

Ex. 6.3.21   Justify (6.3.12), (6.3.20), (6.3.21) and (6.3.26).

Ex. 6.3.22   Let X_n be a Markov chain on S with transition matrix **P**.

1. For $i, j \in S$ show that

$$p_{ij}^n = \sum_{k=1}^{n} f_{ij}^k p_{jj}^{n-k}. \tag{6.3.29}$$

2. If j is transient, then show that $\sum_{n=1}^{\infty} p_{ij}^n < \infty$ for all $i \in S$. (*Hint:* Use the previous part and Proposition 6.3.14.)
3. Suppose S is a finite set. Show that there is at least one recurrent state. (*Hint:* Assume that every state is transient and use the previous part to arrive at a contradiction.)

Ex. 6.3.23   Let X_n be a Markov chain on a state space S with transition matrix P. Let $j \in S$ be transient. Then show that $\sum_{n=1}^{\infty} P_i(X_n = j) = \sum_{m=0}^{\infty} P_i(N(j) \geq m)$ for all $i \in S$. (*Hint:* Use the monotone convergence theorem to conclude that $\sum_{j=1}^{\infty} P_i(X_n = j) = E_i(N(j))$ and then use Ex. 3.1.12.)

6.3.2   Decomposition of a finite state space

Now that we have developed techniques for verifying whether a state is recurrent or transient, we shall focus on a decomposition result when the state space is finite.

Theorem 6.3.24   *Let X_n be a Markov chain on S with transition matrix* **P**. *Suppose S is a finite set. Then S can be written as a disjoint union $T \coprod \left(\coprod_{k=1}^{m} R_k \right)$, where T is the set of all transient states and each $R_k, 1 \leq k \leq m$ is a closed communicating class of recurrent states.*

Proof: Let $T = \{j \in S | \exists k \in S, k \leftarrow j, j \nleftarrow k\}$. By Lemma 6.3.15 (1), all states in T are transient.

As $S \neq T$ (by Ex. 6.3.22), there exists $z_1 \in S \backslash T$. Let $R_1 = \{x \in S : x \leftrightarrow z_1\}$. Note that $R_1 \cap T = \emptyset$. If $S \neq T \coprod R_1$ and $z_2 \in S \backslash (T \coprod R_1)$, let $R_2 = \{x \in S : x \leftrightarrow z_2\}$. Since $|S| < \infty$, this process will end at a finite stage. So there exist $z_1, z_2, \ldots, z_m \in S \backslash T$ such that $R_k = \{x \in S : x \leftrightarrow z_k\}, k = 1, \ldots, m$ are disjoint closed communicating classes. and such that

$$S \backslash T = \coprod_{k=1}^{m} R_k.$$

Consider R_k for some $k \in \{1, 2, \ldots, m\}$. Note that R_k is finite and is a closed communicating class. For any $y \in R_k$, let $n_0 \equiv n_0(y) \in \mathbb{N}$ be such that $p_{yz_k}^{n_0(y)} > 0$. Letting $c = \min_{y \in R_k} p_{yz_k}^{n_0(y)}$, we find that

$$\sum_{y \in R_k} \sum_{m=1}^{\infty} p_{yy}^{m} \geq \sum_{y \in R_k} \sum_{m=n_0(y)+1}^{\infty} p_{yy}^{m}$$

$$\geq \sum_{y \in R_k} \sum_{m=n_0(y)+1}^{\infty} p_{yz_k}^{n_0(y)} p_{z_k y}^{m-n_0(y)} \qquad \text{(By (6.1.5))}$$

$$= \sum_{y \in R_k} p_{yz_k}^{n_0(y)} \sum_{m=1}^{\infty} p_{z_k y}^{m} \geq c \sum_{y \in R_k} \sum_{m=1}^{\infty} p_{z_k y}^{m}$$

$$= c \sum_{m=1}^{\infty} \sum_{y \in R_k} p_{z_k y}^{m} \qquad \text{(By Tonelli's theorem)}$$

$$= c \sum_{m=1}^{\infty} 1 = \infty,$$

where, in the last step, we have used the fact that

$$x \leftarrow y \in R_k \Rightarrow x \leftarrow z_k \Rightarrow x \in R_k.$$

So, by Proposition 6.3.14, we conclude that there is at least one state in each R_k that is recurrent. From Lemma 6.3.15 we conclude that all states in R_k are recurrent.   □

EXERCISES

Ex. 6.3.25   Consider the Markov chain X_n on $S = \{0, 1, 2, 3, 4, 5, 6\}$ with initial distribution μ and transition matrix

$$P = \begin{pmatrix} 0 & \frac{1}{2} & \frac{1}{2} & 0 & 0 & 0 & 0 \\ 0 & \frac{1}{2} & \frac{1}{2} & 0 & 0 & 0 & 0 \\ 0 & \frac{1}{2} & 0 & \frac{1}{2} & 0 & 0 & 0 \\ 0 & 0 & \frac{1}{2} & \frac{1}{2} & 0 & 0 & 0 \\ 0 & 0 & 0 & 0 & 0 & 1 & 0 \\ 0 & 0 & 0 & 0 & 0 & \frac{1}{2} & \frac{1}{2} \end{pmatrix}.$$

Decompose the state space S as in Theorem 6.3.24.

6.4   Strong Markov Property

Let X_n be a Markov chain on S with transition matrix **P**. Many times one is interested in observing the chain at random times (for e.g.: in Example 6.2 define $T = \inf\{n \geq 1 : X_n = 10\}$) and it would be desirable to have the Markov property apply at these random times as well. We shall prove that any discrete time Markov chain has this version—the so-called "strong Markov property." We fix some notations first.

Recall that for $i \in S$, we write E_i for expectation with respect to P_i.

For $n \in \mathbb{N}$, consider the (measurable) shift map $\theta_n : S^W \rightarrow S^W$ defined by $\theta_n(\omega_0, \omega_1, \dots) = (\omega_n, \omega_{n+1}, \dots)$. Then $X_m \circ \theta_n = X_{m+n}$ for all $m, n \in \mathbb{N}$.

Definition 6.4.1   *A random variable $T : \Omega \rightarrow \{0, \infty\} \cup \mathbb{N}$ is said to be a* **stopping time** *for X_n if*

$$\{T = k\} \in \mathcal{B}_k, \tag{6.4.30}$$

for all $k \in \{0, \infty\} \cup \mathbb{N}$ with $\mathcal{B}_n = \sigma\{X_k : k \leq n\}$ and $\mathcal{B}_\infty = \mathcal{B}_W$.

Loosely speaking, a stopping time is a time that can be observed with information up to present time. For instance in Example 6.1, $T = \inf\{n \geq 1 : X_n = 0\}$ is a stopping time but $L = \sup\{n \geq 1 : X_n = 3\}$ is not.

The stopped σ-algebra $\mathcal{B}_T$ and 'random shift' associated to a stopping time T are defined by

$$\mathcal{B}_T = \{A \in \mathcal{B}_W | A \cap \{T = n\} \in \mathcal{B}_n \ \forall n \in \mathbb{N}\} \tag{6.4.31}$$

and

$$\theta_T(\omega) = \theta_{T(\omega)}(\omega).$$

We are now ready to prove the main result of this section.

Theorem 6.4.2   (Strong Markov property) *Let X_n be a Markov chain on S with transition matrix **P**. Let T be a stopping time of X_n and Y be an integrable random variable on $(S^W, \mathcal{B}_W, P)$. Then*

$$E(Y \circ \theta_T | \mathcal{B}_T) = E(Y | X_T) \quad \text{a.e. on } \{T < \infty\}[1],$$

where, of course, $X_T(\omega) = X_{T(\omega)}(\omega)$.

---

[1]Two random variables W and Z on a probability $(\Omega, \mathcal{B}, P)$ are equal a.e. on $B \in \mathcal{B}$ if $W 1_B = Z 1_B$ a.e.

Proof: By Ex. 6.4.4, we may find a Borel-measurable function $\phi : S \to \mathbb{R}$ such that ϕ is Borel measurable such that $E(Y|X_0) = \phi(X_0)$. For $A \in \mathcal{B}_T$ we have,

$$\int_A 1_{\{T<\infty\}} E(Y \circ \theta_T | \mathcal{B}_T) dP = E(E((Y \circ \theta_T) 1_A 1_{\{T<\infty\}} | \mathcal{B}_T))$$

$$= E((Y \circ \theta_T) 1_A 1_{\{T<\infty\}})$$

$$= \sum_{n=1}^{\infty} E(Y \circ \theta_n 1_A 1_{\{T=n\}})$$

$$= \sum_{n=1}^{\infty} E(E(Y \circ \theta_n | \mathcal{B}_n) 1_{A \cap \{T=n\}})$$

$$= \sum_{n=1}^{\infty} E(\phi(X_n) 1_{A \cap \{T=n\}})) \text{ (by Ex. 6.4.5)}$$

$$= \int_A \phi(X_T) 1_{\{T<\infty\}} dP.$$

$\square$

We present two applications of the strong Markov property. Define the so-called **kth return time** to $i \in S$ (of a Markov chain on state space S) by

$$U_i^k = \inf\{n \geq 1 : N_n(i) = k\}.$$

and the **inter-arrival times** at $i \in S$ for the chain X_n by

$$I_i^k = U_i^k - U_i^{k-1}$$

for $k \geq 2$.

Lemma 6.4.3   *Let X_n be a Markov chain on S with transition matrix* **P**. *Assume that the chain is irreducible and $i \in S$ is reccurent. With U_i^1, $\{I_i^k : k \geq 2\}$ as above, we have:*

(a) $P_j(U_i^1 < \infty) = 1$, *for all $j \in S$; and*
(b) $\{I_i^k : k \geq 2\}$ *is an i.i.d. sequence of random variables with mean $E_i(T_i)$.*

Proof:

(a)   is immediate from Ex. 6.3.11.

(b)   Let $k \geq 1$. From Ex. 6.3.12, one sees that the 'inf' in the definition of U_i^k is indeed a 'min' a.e. So $X_{U_i^1} = i$ a.e.

We shall first prove by induction that $X_{U_i^m} = i$ a.e. Thus, assume this is true for $m \leq k$. Then, by the inductive hypothesis and strong Markov property we have

$$P(X_{U_i^{k+1}} = i) = E(1_{\{X_{U_i^{k+1}}=i\}})$$

$$= E(E(1_{\{X_{U_i^1}=i\}} \circ \theta_{U_i^k} \mid \mathcal{B}_{U_i^k}))$$

$$= E(\psi(X_{U_i^k})) = \psi(i) = P_i(X_{U_i^1} = i) = 1,$$

where ψ is the ϕ from Ex. 6.4.4 (with $W = 1_{\{X_{U_i^1}=i\}}$, $V = X_{U_i^{k-1}}$). So, indeed

$$X_{U_i^m} = i \text{ a.e. for all } m \in \mathbb{N} \tag{6.4.32}$$

For $i \in S, k \geq 2$, and $m \in \mathbb{N}$, by the strong Markov property we have,

$$P(I_i^k = m) = E(E(1_{\{U_i^k - U_i^{k-1}=m\}} \mid \mathcal{B}_{U_i^{k-1}}))$$

$$= E(E(1_{\{U_i^1=m\}} \circ \theta_{U_i^{k-1}} \mid \mathcal{B}_{U_i^{k-1}}))$$

$$= E(\eta(X_{U_i^{k-1}})) = \eta(i) = P_i(U_i^1 = m),$$

where η is the ϕ from Ex. 6.4.4 (with $W = 1_{\{U_i^1=m\}}$, $V = X_{U_i^{k-1}}$) and we have used (6.4.32) in the last step. So

$$I_i^k \overset{d}{=} U_i^1, \tag{6.4.33}$$

for all $k \geq 2$ and $i \in S$.

For $2 \leq k < l$, and $m, n \in \mathbb{N}$, by the strong Markov property we have,

$$P(I_i^k = m, I_i^l = n) = E(E(1_{I_i^k=m} 1_{\{U_i^l - U_i^{l-1}=n\}} \mid \mathcal{B}_{U_i^{l-1}}))$$

$$= E(1_{I_i^k=m} E(1_{U_i^1=n} \circ \theta_{U_i^{l-1}} \mid \mathcal{B}_{U_i^{l-1}}))$$

$$= E(1_{I_i^k=m} \zeta(X_{U_i^{l-1}})) = \zeta(i) E(1_{I_i^k=m})$$

$$= P_i(U_i^1 = n) P(I_i^k = m)$$

$$= P(I_i^k = m) P(I_i^l = n),$$

where ζ is the ϕ from Ex. 6.4.4 (with $W = 1_{U_i^1=n}$, $V = X_{U_i^{l-1}}$) and we have used (6.4.32) in the third last step and (6.4.33) in the last step.
So I_i^k are i.i.d $\mathbb{N}$-valued random variables and $E(I_i^k) = E_i(U_i^1) = E_i(T_i)$. $\square$

Secondly we try to understand some interesting questions that arise in Example 6.1.

Example 6.7 **(One-step recursion)** Let X_n be a Markov chain on state space $S = \{0, 1, 2, \ldots, N\}$ with initial distribution μ and transition matrix $\mathbf{P}$ given by:

$$p_{ij} = \begin{cases} 1 & \text{if } i = 0, j = 0 \text{ or } i = N, j = N \\ \frac{1}{2} & \text{if } i = j + 1, 1 \leq i \leq N - 1 \\ \frac{1}{2} & \text{if } i = j - 1, 1 \leq i \leq N - 1 \\ 0 & \text{otherwise.} \end{cases}$$

Note that this is a generalised version of the Gambler's ruin chain discussed in Example 6.1, where the total capital available is N instead of 4. There are two quantities of interest for this Markov chain.

1. The "expected duration of the game", i.e. $E_i(D)$, where $i \in S$, $D = \min\{n \geq 0 : X_n \in \{0, N\}\}$.
    Let $f : S \to \mathbb{R}$ be defined as $f(i) = E_i(D)$. Now,

    $$f(0) = f(N) = 0. \tag{6.4.34}$$

    and using the symmetry of the chain, $f(i) = f(N - i)$ for $i \in S$. The strong Markov property implies that for $i \in S, i \neq 0, N$

    $$f(i) = 1 + \frac{1}{2}f(i + 1) + \frac{1}{2}f(i - 1). \quad \text{(justify)} \tag{6.4.35}$$

    Using the above three observations one can solve for f to get

    $$f(i) = i(N - i) \text{ for } i \in S.$$

2. The probability of Pyare Lal being the winner is

    $$P_i(T_N < T_0), \text{ for } i \in S.$$

    Let $g : S \to \mathbb{R}$ be defined as $g(i) = P_i(T_N < T_0)$ Now,

    $$g(0) = 0, g(N) = 1 \tag{6.4.36}$$

    and the strong Markov property implies (justify) that for $i \in S, i \neq 0, N$

    $$g(i) = \frac{1}{2}g(i + 1) + \frac{1}{2}g(i - 1). \quad \text{(justify)} \tag{6.4.37}$$

    Using the above two observations one can solve for g to get $g(i) = \frac{i}{N}$ for $i \in S$.

Thus both the quantities of interest have explicit formulae. The technique used to obtain these formulae is usually referred to as obtaining the 'one step recursion formula'.

EXERCISES

Ex. 6.4.4 Suppose V, W are two S-valued random variables on a probability space $(\Omega, \mathcal{B}, P)$. If $E(W|V)$ is defined to be the conditional expectation of W given $\sigma(V)$ (i.e. the σ-algebra generated by V), show that there is a measurable $\phi : S \to \mathbb{R}$ such that $E(W|V) = \phi(V)$ a.e.

Ex. 6.4.5 Let X_n be a Markov chain on S with transition matrix $\mathbf{P}$. Let $Y :$ $S \to \mathbb{R}$ be $(S^W, \mathcal{B}^W)$ measurable. Let $\phi : S \to \mathbb{R}$ (from Ex. 6.4.4) be such that $E(Y \mid X_0) = \phi(X_0)$ a.e. Then show that

$$E(Y \circ \theta_n | \mathcal{B}_n) = \phi(X_n) \quad a.e. \tag{6.4.38}$$

(*Hint:* It is enough to consider $Y = 1_B$, $B \in \mathcal{B}_W$; in that case, use the definition of conditional expectation.)

Ex. 6.4.6 Consider the Markov chain in Example 6.2. Show that $L = \sup\{n \geq 1 : X_n = 1\}$ is not a stopping time.

Ex. 6.4.7 Consider the Markov chain in Example 6.1. Show that $F = \inf\{n \geq 1 : X_n \in \{0, 4\}\}$ is a stopping time. Can you find the distribution of X_F ?

Ex. 6.4.8 Let (X_n, P) be a Markov chain on a state space S. Show that the return time T_i to a state $i \in S$ is a stopping time.

Ex. 6.4.9 Consider the Markov chain (X_n, P) discussed in Example 6.2. Let $X_0 = 0$ a.e. Let $a > 0$. Let

$$T_a = \inf\{n \geq 1 : X_n = a\}.$$

(a) Show that T_a is a stopping time and the 'inf' in T_a is actually a minimum almost everywhere.

(b) (**Reflection principle**) Suppose $M_n = \max_{0 \leq i \leq n} X_i$. Show that

$$P(M_n \geq a, X_n < a) = P(M_n \geq a, X_n > a).$$

(*Hint:* Apply the strong Markov property at T_a and symmetry of the distribution of the Bernoulli trials.)

Ex. 6.4.10   Justify (6.4.35)and (6.4.37). Show that $f(i) = i(N - i)$ and $g(i) = \frac{i}{N}$ for $i \in S$.

Ex. 6.4.11   Provide a proof of (6.3.12) using the strong Markov property.

6.5   Stationary Distribution

Let X_n be a Markov chain on S with transition matrix **P**.

Definition 6.5.1   *A probability measure π on S is said to be stationary for the Markov chain X_n or its* **Stationary Distribution** *if*

$$\sum_{i \in S} \pi(i) p_{ij} = \pi(j). \tag{6.5.39}$$

As the definition indicates, if the chain were to start with initial distribution π, then its distribution always remains at π. This is made precise below.

Proposition 6.5.2   *Let X_n be a Markov chain with transition matrix P on state space S.*

1. *Let $n \in \mathbb{N}$. Suppose π is a stationary distribution of X_n, then $X_0 \stackrel{d}{=} \pi$ implies $X_n \stackrel{d}{=} \pi$.*
2. *If $|S| < \infty$ and P is doubly stochastic, $\left(i.e\ also\ \sum_{i \in S} p_{ij} = 1\ for\ all\ j \in S\right)$, then a stationary distribution for X_n is given by $\pi(i) = \frac{1}{|S|}$ for all $i \in S$.*

Proof:

(1)   As π is stationary, this is immediate from (6.5.1) and definition of a Markov chain.

(2)   If $|S| = N$ and $\pi(i) = \frac{1}{N}$ for all $i \in S$, then, for each $j \in S$, observe that

$$\sum_{i \in S} \pi(i) p_{ij} = \frac{1}{N} \sum_{i \in S} p_{ij} = \frac{1}{N} = \pi(j).$$

Also $\sum_{i \in S} \pi(i) = 1$. So π is a stationary distribution of X_n.   □

Example 6.8   (**Birth–death chains**) Let Let X_n be a Markov chain on the state space $S = \{l, l+1, \dots, u-1, u\}$ with $-\infty < l < u < \infty$ and 'tridiagonal' probability transition matrix given by

$$P = \begin{bmatrix} z_l & p_l & 0 & \cdots & \cdots & 0 \\ m_{l+1} & z_{l+1} & p_{l+1} & \ddots & \vdots & 0 \\ \vdots & \ddots & \ddots & \ddots & \ddots & \vdots \\ 0 & \ddots & \ddots & m_{u-1} & z_{u-1} & p_{u-1} \\ 0 & \cdots & \cdots & 0 & m_u & z_u \end{bmatrix}.$$

Assume that m_i and p_i are strictly positive where defined. If the chain has a stationary distribution π, then it is not hard to see that

$$\pi(l+i) = c \frac{p_{l+i-1} p_{l+i-2} \cdots p_l}{m_{l+i} m_{l+i-1} \cdots m_{l+1}}.$$

The fact that π is a probability on S will imply that

$$\pi(l) = c = \frac{1}{1 + \sum_{k=l+1}^{u} \frac{p_{k-1} p_{k-2} \cdots p_l}{m_k m_{k-1} \cdots m_{l+1}}}.$$

These chains, where $S \subset \mathbb{Z}$ and $p_{ij} = 0$ if $|i - j| > 1$, are known as birth–death chains. These are used to model biological systems.

While not every Markov chain may have a stationary distribution, it is nevertheless true that every irreducible Markov chain on a finite state space S has one, which is further unique, as is established in Remark 6.6.3 (2)). Our proof below is borrowed from [13].

Theorem 6.5.3 *Let X_n be an irreducible Markov chain on a finite state space S. Then X_n has a stationary distribution.*

Proof: For $i \in S$, a directed spanning tree on S with root at i is by definition a graph g which satisfies the following conditions:

(a) g has vertices indexed by S, and is a tree, meaning that it is connected and has no cycles; or equivalently, there is a unique path between any two vertices in g.

(b) The edges of g are directed in such a way that for any $j \in S$, all the edges in the unique path from j to i is directed towards i, i.e., the path from j to i is directed towards i.

Let $G(i)$ be the set of all directed spanning trees on S with root at i. For every $i \in S$, define

$$\pi_0(i) = \sum_{g \in G(i)} \prod_{(j \to k) \in g} p_{jk}, \tag{6.5.40}$$

where $j \to k$ denotes any directed edge in the tree g. As X_n is irreducible each $\pi_0(i) > 0$. We shall verify that π_0 stationary distribution, i.e.

$$\pi_0(i) = \sum_{j \in S} \pi_0(j) p_{ji}, \forall i \in S,$$

which is equivalent to showing that

$$\pi_0(i) \sum_{j \neq i} p_{ij} = \sum_{j \neq i} \pi_0(j) p_{ji} \qquad (6.5.41)$$

or that

$$\left(\sum_{g \in G(i)} \prod_{(j \to k) \in g} p_{jk} \right) \sum_{j \neq i} p_{ij} = \sum_{j \neq i} \left(\sum_{g \in G(j)} \prod_{(k \to j) \in g} p_{kj} \right) p_{ji} \qquad (6.5.42)$$

The equality follows from the following observation, which can be easily verified. Both sides of (6.5.42) are equal to $\sum_{g \in L} \prod_{(k \to j) \in g} p_{jk}$, where L is the set of all graphs satisfying:

(a) every point $i \in S$ has only one directed edge pointing way from i to $j \in S$, $j \neq i$.

(b) the graph has exactly one closed cycle and $i \in S$ belongs to that cycle.

So the normalised version of π_0, defined by $\pi(i) = \left(\sum_{j \in S} \pi_0(j) \right)^{-1} \pi_0(i)$, will be a stationary distribution for X_n. $\square$

EXERCISES

Ex. 6.5.4 Let X_n be a Markov chain on $S = \{1, 2\}$ with initial distribution μ and transition matrix

$$P = \begin{pmatrix} 1 - \alpha & \alpha \\ \beta & 1 - \beta \end{pmatrix},$$

where $\alpha, \beta \in [0, 1]$ but $\alpha + \beta > 0$.

1. Find the stationary distribution π.
2. Find the eigenvalues (and the corresponding eigenvectors) of P.
3. From the previous step calculate P^n and $\lim_{n \to \infty} P^n$. What can you say about the limiting matrix ?

Ex. 6.5.5 Consider the Markov chain with state space $S = \{0, 1, 2, \dots\}$, with the transition matrix P given by:

$$p_{ij} = \begin{cases} 1 - p_i & \text{if } j = i + 1 \\ p_i & \text{if } j = 0. \end{cases}$$

Specify conditions on p_i under which the chain: (a) will be recurrent; and (b) will have a stationary distribution ?

6.6   Limit Theorems

Limiting behaviour is of importance in many applications. That is, as $n \to \infty$,

(i)   does a Markov chain approach a limiting distribution ?

(ii)   If so, does this limiting distribution depend on the distribution of X_0 ?

In this section we prove the two main limit theorems for Markov chains. The first one is sometimes referred to as the Renewal Theorem.

Theorem 6.6.1   *Let X_n be an irreducible Markov chain on S with transition matrix P. For $i \in S$*

$$\frac{N_n(i)}{n} \xrightarrow{a.e.} \frac{1}{E_i(T_i)}. \tag{6.6.43}$$

The second result is the main convergence result in the area of Markov chains.

Theorem 6.6.2   *Let X_n be a Markov chain on S with transition matrix $\mathbf{P}$. Assume that X_n is aperiodic and irreducible. If X_n has a stationary distribution π and if $P(X_0 = i) > 0$, then*

$$\lim_{n \to \infty} p_{ij}^n = \pi(j), \tag{6.6.44}$$

for all $j \in S$.

Before proving the above two results, we point out some of their implications and possible generalisations.

Remark 6.6.3

1.   *Generalisations:*

a)   *The irreducibility assumption in Theorem 6.6.1 is not really necessary. If the irreducibility hypothesis was dropped, then the right-hand side of (6.6.43) should be replaced by $\frac{1_{\{T_i < \infty\}}}{E_i(T_i)}$.*

b)   *If the chain in Theorem 6.6.2 was periodic with period $d > 1$, it can be shown that there exists an integer r with $0 \le r < d$ such that*

$$\lim_{m \to \infty} p_{ij}^{md+r} = d\pi(j) \qquad (6.6.45)$$

for all $j \in S$.

Reason: Part (a) is easily seen from the proof of Theorem 6.6.1. See Ex. 6.6.9 for proof of part (b).

2. *Let X_n be a Markov chain on S with transition matrix $\mathbf{P}$. Suppose X_n is a irreducible. It can have at most one stationary distribution π.*
   Reason: This can be reasoned as follows. Assume that X_n has two stationary distributions π_1, π_2. Consider a Markov chain $\tilde{X}_n$ on state space S with transition matrix $\tilde{\mathbf{P}} = \frac{I+\mathbf{P}}{2}$ where I is the identity matrix. Note that $\tilde{X}_n$ is an aperiodic and irreducible Markov chain with stationary distributions π_1 and π_2 (verify this). Now Theorem 6.6.2 applied to the chain $\tilde{X}_n$ implies that $\pi_1 = \pi_2$.

3. **(Ergodic theorem)** *Let X_n be a Markov chain on S with transition matrix $\mathbf{P}$. Assume that X_n is irreducible and aperiodic. For $i, j \in S$, we have*

$$\frac{\sum_{k=1}^{n} p_{ij}^k}{n} \to \pi(j) \quad a.e. \qquad (6.6.46)$$

This is known as the ergodic theorem for Markov chains.
Reason: This is immediate from Theorem 6.6.2. See Ex. 6.6.8 for an alternate proof.

4. *Let X_n be a Markov chain on S with transition matrix $\mathbf{P}$. If X_n is irreducible, aperiodic and equipped with a stationary distribution π, then*

$$\pi(j) = \frac{1}{E_j(T_j)} \qquad (6.6.47)$$

for all $j \in S$.
Reason : Observe that

$$\sum_{i \in S} \sum_{k=1}^{n} p_{ij}^k \mu(i) = E(N_n(j)). \qquad (6.6.48)$$

Then (6.6.47) is an easy consequence of dominated convergence, Theorem 6.6.1 and (6.6.46).

Proof of Theorem 6.6.1:

Case 1: i is transient.

As $N_n(i) < N(i) < \infty$ a.e. and $E_i(T_i) = \infty$, the result follows.

Case 2: i is recurrent and $E_i(T_i) < \infty$.

Let $k \in \mathbb{N}$, define $U_i^k = \min\{n \geq 1 : N_n(i) = k\}$, the so called the kth return time to i. Let $I_i^k = U_i^k - U_i^{k-1}$ for $k \geq 2$, denote the inter-arrival times at $i \in S$ for the chain X_n. By Lemma 6.4.3 we have that $\{I_i^k : k \geq 2\}$ is an i.i.d. sequence and $E(I_i^k) = E_i(T_i) < \infty$. Hence by the strong law of large numbers (5.4.26), we have

$$E_i(T_i) = \lim_{n \to \infty} \frac{\sum_{k=2}^{n+1} I_i^k}{n} = \lim_{n \to \infty} \frac{U_i^{n+1} - U_i^1}{n} = \lim_{n \to \infty} \frac{U_i^n}{n} \quad \text{a.e.,}$$

where the last equality is due to the fact that $P_j(U_i^1 < \infty) = 1$ for all $j \in S$ (by Lemma 6.4.3). As $i \in S$ is recurrent, $N_n(i)$ increases to ∞ a.e. Therefore

$$\frac{U_i^{N_n(i)}}{N_n(i)} \to E_i(T_i) \quad \text{a.e.}$$

Also note that for any $n \in \mathbb{N}$, $U_i^{N_n(i)} \leq n < U_i^{N_n(i)+1}$. Therefore

$$\frac{U_i^{N_n(i)}}{N_n(i)} \leq \frac{n}{N_n(i)} \leq \frac{U_i^{N_n(i)+1}}{N_n(i) + 1} \frac{N_n(i) + 1}{N_n(i)}.$$

Let $n \to \infty$ in the above inequality to obtain the result.

Case 3: i is recurrent and $E_i(T_i) = \infty$.

For $k \in \mathbb{N}$, $i \in S$, let U_i^k and I_i^k be the random variables defined in Case 2. As $P(T_i < \infty | X_0 = i) = 1$ and the chain is irreducible, this implies $P_j(U_i^1 < \infty) = 1$ for all $j \in S$. Also, as above, I_i^k are non-negative i.i.d. random variables such that $E(I_i^k) = \infty$. Now, we appeal to Ex. 6.6.6 to conclude that

$$\lim_{n \to \infty} \frac{\sum_{k=2}^{n+1} I_i^k}{n} = \infty \text{ a.e.}$$

The proof then follows as in Case 2.   □

Using the proof of Theorem 6.6.1 as a cue, the recurrent states can be classified further depending on whether the return times have finite mean or not. More precisely,

Definition 6.6.4   *Let X_n be a Markov chain on S with transition matrix* **P**. *Let i be a recurrent state. Then i is said to be*

1.   **positive recurrent** *if $E_i(T_i) < \infty$.*

2.   **null recurrent** *if $E_i(T_i) = \infty$.*

A chain is said to be positive (null) recurrent if every state is positive (null) recurrent.

Proof of Theorem 6.6.2: Let $S^2 = S \times S$. Let Y_n be a Markov chain on S, as in Remark 6.3.16, which is independent of X_n, and has initial distribution π and the same transition matrix $\mathbf{P}$. Let $Z_n = (X_n, Y_n)$.

It is easy to verify that Z_n is a Markov Chain on $S \times S$ with initial distribution $\mu \times \pi$ and transition matrix $\widetilde{\mathbf{P}}$ given by

$$\widetilde{p}_{\mathbf{i}_1 \mathbf{i}_2} = p_{i_1 i_2} p_{j_1 j_2}, \tag{6.6.49}$$

where $\mathbf{i}_k = (i_k, j_k) \in S^2$, for $k \in \{1, 2\}$.

Fix $i_1, i_2, j_1, j_2 \in S$. Since X_n is irreducible, there exist $r, s \in \mathbb{N}$ such that $p^r_{i_1 i_2} > 0$ and $p^s_{j_1 j_2} > 0$. As X_n is aperiodic, by Ex. 6.3.13, we know that there is a $t_0 \in \mathbb{N}$ such that for all $t \geq t_0$, $p^{s+t}_{i_2 i_2}$ and $p^{r+t}_{j_2 j_2}$ are strictly positive. Therefore for $t \geq t_0$, we have shown that

$$\widetilde{p}^{\,r+s+t}_{\mathbf{i}_1 \mathbf{i}_2} = p^{r+s+t}_{i_1 i_2} p^{r+s+t}_{j_1 j_2}$$

$$\geq p^r_{i_1 i_2} p^{s+t}_{i_2 i_2} p^s_{j_1 j_2} p^{r+t}_{j_2 j_2}$$

$$> 0$$

Since $\mathbf{i}_1, \mathbf{i}_2$ were arbitrary, we have shown that Z_n is irreducible. Define $\pi(\mathbf{i}) = \pi(i)\pi(j)$ for any $\mathbf{i} = (i, j) \in S^2$. It is easily verified that $\pi(\cdot)$ is stationary distribution for Z_n. Hence by Ex. 6.6.7, Z_n is recurrent.

Let $T = \min\{n \geq 1 : X_n = Y_n\}$ and $T_d = \min\{n \geq 1 : X_n = Y_n = d\}$. As Z_n is recurrent, $P(T_d < \infty) = 1$. As $T \leq T_d$, $P(T < \infty) = 1$. We will complete the proof of (6.6.44) assuming the following key coupling inequality whose proof we shall subsequently give:

$$|P(X_n = j, X_0 = i) - P(Y_n = j, X_0 = i)| \leq P(T > n). \tag{6.6.50}$$

We shall assume henceforth that $Y_0 \overset{d}{=} \pi$ (and hence $Y_n \overset{d}{=} \pi$ by Proposition 6.5.2). Assuming (6.6.50), we have,

$$\limsup_{n \to \infty} |p^n_{ij} - \pi(j)| = \limsup_{n \to \infty} |P(X_n = j | X_0 = i) - \pi(j)|$$

$$= \limsup_{n \to \infty} |P(X_n = j | X_0 = i) - P(Y_n = j)|$$

$$= \limsup_{n \to \infty} \frac{|P(X_n = j, X_0 = i) - P(Y_n = j, X_0 = i)|}{\mu(i)}$$

$$\leq \frac{1}{\mu(i)} \limsup_{n\to\infty} P(T > n) \qquad \text{(by (6.6.50))}$$

$$= 0 \qquad \text{(as } P(T < \infty) = 1).$$

Hence $\lim_{n\to\infty} |p_{ij}^n - \pi(j)| = 0$.

We now establish (6.6.50).

$$P(X_n = j, X_0 = i, T \leq n) = E(1_{\{X_0=i,T\leq n\}} E(1_{\{X_n=j\}}|\mathcal{B}_T^Z))$$

$$= E(1_{\{X_0=i,T\leq n\}} \phi^X(Z_T)) \qquad (6.6.51)$$

where ϕ^X is the ϕ from Ex. 6.4.4 (with $W = 1_{\{X_{n-T}=j\}}$, $V = Z_T$); we have used the strong Markov property in the last equality. Similarly,

$$P(Y_n = j, X_0 = i, T \leq n) = E(1_{\{X_0=i,T\leq n\}} a\phi^Y(Z_T)), \qquad (6.6.52)$$

where ϕ^Y is the ϕ from Ex. 6.4.4 (with $W = 1_{\{Y_{n-T}=j\}}$, $V = Z_T$). A . $X_T = Y_T$ a.e., we have

$$\phi^X(Z_T) = \phi^Y(Z_T). \quad \text{a.e. (why ?).} \qquad (6.6.53)$$

Therefore, from (6.6.53), (6.6.51) and (6.6.52), we have

$$P(X_n = j, X_0 = i) = P(X_n = j, X_0 = i, T \leq n)$$

$$+ P(X_n = j, X_0 = i, T > n)$$

$$= P(Y_n = j, X_0 = i, T \leq n)$$

$$+ P(X_n = j, X_0 = i, T > n)$$

$$\leq P(Y_n = j, X_0 = i) + P(T > n).$$

As the last inequality is symmetric in X_n and Y_n, we have established (6.6.50). □

EXERCISES

Ex. 6.6.5 Prove (6.6.53).

Ex. 6.6.6 Let $X_n, n \geq 1$ be i.i.d. non-negative random variables such that $E(X_1) = \infty$. Let $S_n = \sum_{k=1}^n X_k$. Then show that $\sum_{n=1}^\infty \frac{S_n}{n} \to \infty$.

Ex. 6.6.7 Let X_n be a Markov chain on S with transition matrix $\mathbf{P}$. Suppose that X_n is irreducible and has a stationary distribution π. Show that the chain is recurrent. (*Hint:* Note that there is an $i \in S$ such that $\pi(i) > 0$. Show that

$$\pi(i) = \sum_{j \in S} \pi(j) \frac{\sum_{k=1}^{n} p_{ji}^{k}}{n}.$$

Then use Theorem 6.6.1 and dominated convergence to conclude that $i \in S$ is recurrent. Irreducibility and Lemma 6.3.15 imply the result.)

Ex. 6.6.8   Provide an alternate proof of (6.6.46) using Theorem 5.4.4.

Ex. 6.6.9   Prove Remark 6.6.3 1(b).
(*Hint:* Define $Y_n = X_{nd}$ for $n \geq 1$. Then Y_n is a Markov chain on S with initial distribution μ and transition matrix $Q = P^d$. Observe that Y_n is aperiodic. Let $j \in S$, $A_j = \{i \in S : j \leftrightarrow i\}$ (accesibility for Y_n). Show that

$$q_{ij}^{m} \to \frac{1}{E(T_j^Y)} = d \frac{1}{E_j(T_j)} = d\pi(j),$$

for all $i \in A_j$. Let $r = \inf\{n \geq 1 : p_{ij}^n > 0\}$. As

$$p_{ij}^{md+r} = \sum_{k=1}^{n} f_i^{kd+r} {}_{j} p_{jj}^{(m-k)d} = \sum_{k=1}^{n} f_i^{kd+r} {}_{j} p_{jj}^{(m-k)d},$$

let $m \to \infty$ and complete the proof.)

Ex. 6.6.10   Let X_n be a Markov chain on S with transition matrix **P**. Let $i \in S$ be a positive recurrent state. Let $\nu : S \to [0, \infty)$ by

$$\mu(j) = \sum_{k=0}^{\infty} P_i(X_k = j, T_i > k)$$

for $j \in S$. Show that $\pi : S \to [0, 1]$ given by

$$\pi(j) = \frac{\nu(j)}{E_i(T_i)}$$

for $j \in S$ defines a stationary distribution for the chain X_n.

Ex. 6.6.11   Let X_n be a Markov chain on S with transition matrix **P**. Suppose $i \in S$ is a positive recurrent state and i communicates with j, then $j \in S$ is also positive recurrent.

7

Some Analysis

7.1 Complex Measures

It will be necessary, for a variety of reasons, to consider measures which are complex-valued. Thus, a complex measure on a measurable space $(\Omega, \mathcal{B})$ is nothing but a function $\mu : \mathcal{B} \to \mathbb{C}$ which is countably additive – meaning, of course, that if $E = \bigsqcup_{n=1}^{\infty} E_n$ is a measurable partition of a set $E \in \mathcal{B}$, then the complex series $\sum_{n=1}^{\infty} \mu(E_n)$ is convergent and converges to $\mu(E)$. There is essentially one basic fact about such complex measures which this section is devoted to establishing, that such a measure is expressible in the form $(\mu_1 - \mu_2) + i(\mu_3 - \mu_4)$, where the μ_i's are *finite* positive measures.

To begin with, we may consider the real and imaginary parts of μ denoted by $\text{Re}\,\mu$ and $\text{Im}\,\mu$ respectively, defined as

$$(\text{Re}\,\mu)(E) = \text{Re}(\mu(E)), \quad (\text{Im}\,\mu)(E) = \text{Im}(\mu(E)).$$

It should be clear that $\text{Re}\,\mu$ and $\text{Im}\,\mu$ are real-valued and countably additive set-functions (which do NOT assume infinite values on any sets). We shall refer to such 'measures' as **finite real measures** and to their complex analogues as **finite complex measures**.

Until Corollary 7.1.6, the symbol μ will denote a fixed 'finite real measure' as above. We begin with a lemma.

Lemma 7.1.1 *If $\{E_n\}$ is any sequence of pairwise disjoint $\mathcal{B}$-measurable sets, then*

$$\sum_{n=1}^{\infty} |\mu(E_n)| < \infty. \tag{7.1.1}$$

Proof: Define

$$A_n = \begin{cases} E_n & \text{if } \mu(E_n) \geq 0 \\ \emptyset & \text{if } \mu(E_n) \leq 0 \end{cases}$$

and

$$B_n = \begin{cases} E_n & \text{if } \mu(E_n) \leq 0 \\ \emptyset & \text{if } \mu(E_n) \geq 0 \end{cases}$$

and note that $\{A_n : n \in \mathbb{N}\}$ and $\{B_n : n \in \mathbb{N}\}$ are sequences of pairwise disjoint $\mathcal{B}$-measurable sets, and that $E_n = A_n \bigsqcup B_n$. So,

$$\sum_{n=1}^{\infty} |\mu(E_n)| = \left(\sum_{n=1}^{\infty} \mu(A_n) \right) + \left| \left(\sum_{n=1}^{\infty} \mu(B_n) \right) \right|$$

$$= \mu \left(\bigsqcup_n A_n \right) + \left| \mu \left(\bigsqcup_n B_n \right) \right|$$

$$< \infty .$$ □

We now come to a crucial definition.

Definition 7.1.2   *With μ as above, define the set function $|\mu| : \mathcal{B} \to [0, \infty]$ by*

$$|\mu|(E) = \sup \left\{ \sum_{n=1}^{\infty} |\mu(E_n)| : E = \bigsqcup_{n=1}^{\infty} E_n , \ E_n \in \mathcal{B} \ \forall n \right\} . \qquad (7.1.2)$$

*Then $|\mu|$ is called the **total variation measure** associated with μ.*

We begin with a simple proposition that justifies the above terminology.

Lemma 7.1.3   *With the foregoing notation, $|\mu|$ is a (countably additive) measure.*

Proof: Since $|\mu|(\emptyset) = 0$ by definition, we only need to check the countable additivity. So, suppose $F = \bigsqcup_{m=1}^{\infty} F_m$ where $F_m \in \mathcal{B} \ \forall m$. On the one hand, if $F = \bigsqcup_{n=1}^{\infty} E_n$ with $E_n \in \mathcal{B} \ \forall n$, then we have

$$\sum_{n=1}^{\infty} |\mu(E_n)| = \sum_{n=1}^{\infty} \left| \sum_{m=1}^{\infty} \mu(E_n \cap F_m) \right|$$

$$\leq \sum_{n=1}^{\infty} \sum_{m=1}^{\infty} |\mu(E_n \cap F_m)|$$

$$= \sum_{m=1}^{\infty} \sum_{n=1}^{\infty} |\mu(E_n \cap F_m)|$$

$$\leq \sum_{m=1}^{\infty} |\mu|(F_m);$$

by taking the supremum over the partition $\coprod_n E_n$ we find that

$$|\mu|(F) \le \sum_{m=1}^{\infty} |\mu|(F_m).$$

On the other hand, let $F_m = \coprod_{n=1}^{\infty} F_{m,n}$ be an arbitrary partition of F_m for each m. Then $\coprod_{m,n=1}^{\infty} F_{m,n}$ is a partition of F and consequently, we find that

$$|\mu|(F) \ge \sum_{m,n=1}^{\infty} |\mu(F_{m,n})|$$

$$= \sum_{m=1}^{\infty} \sum_{n=1}^{\infty} |\mu(F_{m,n})|;$$

by varying the partitions $\{F_{m,n} : n = 1, 2, \dots\}, m = 1, 2, \dots$, we find that

$$|\mu|(F) \ge \sum_{m=1}^{\infty} |\mu|(F_m).$$

$\square$

We are now ready for the main result of this section.

Proposition 7.1.4   *With the foregoing notation, $|\mu|$ is a **finite** measure.*

Proof: Suppose, if possible, that $|\mu|(\Omega) = \infty$. We will arrive at the desired contradiction after going through the following sequence of steps. (Thus, we assume, throughout this proof, that the proposition is false.)

(1)   If $E \in \mathcal{B}$ and if $|\mu|(E) = \infty$, then there exists a partition $E = A \coprod B$, $A, B \in \mathcal{B}$ such that $|\mu|(A) > 0$ and $|\mu|(B) > 0$. (In other words, the measure $|\mu|$ has no atom of infinite measure.)

Proof of (1):   Suppose, to the contrary, that E is an atom of infinite measure—meaning that (i) $|\mu|(E) = \infty$ and (ii) whenever we have a measurable partition $E = A \coprod B$ of E, then either $|\mu|(A) = 0$ or $|\mu|(B) = 0$. Note that the condition $|\mu|(A) = 0$ implies that $\mu(C) = 0$ for every measurable subset C of A. Thus the 'atom' assumption implies that if $E = \coprod_{n=1}^{\infty} E_n$ is any measurable partition of E, then $\mu(E_n) = 0$ for all but one n, say $n = n_0$, and consequently that $\mu(E) = \mu(E_{n_0})$ by countable additivity of μ. Hence,

$$\sum_{n=1}^{\infty} |\mu(E_n)| = |\mu(E)|$$

for any measurable partition $\{E_n\}$ of E. Thus, by definition of $|\mu|$, we find that $|\mu|(E) = |\mu(E)| < \infty$, contradicting the assumption that $|\mu|(E) = \infty$. This contradiction completes the proof.

(2)   Suppose $F \in \mathcal{B}$ and $|\mu|(F) = \infty$. Define $\mathcal{C}_F = \{E \subset F : E \in \mathcal{B}, |\mu| (F \setminus E) = \infty\}$, and $\alpha = \sup\{|\mu|(E) : E \in \mathcal{C}_F\}$. Then, $\alpha = \infty$.

Proof of (2): Suppose, conversely, that $\alpha < \infty$.
The first thing to notice is that $\mathcal{C}_F$ is 'hereditary' in the sense that

$$E_1, E_2 \in \mathcal{B}, \ E_2 \in \mathcal{C}_F, \ E_1 \subset E_2 \ \Rightarrow \ E_2 \in \mathcal{C}_F. \tag{7.1.3}$$

This has the pleasant consequence that $\mathcal{C}_F$ is closed under (finite) unions. Indeed, suppose $E_1, E_2 \in \mathcal{C}_F$, then (7.1.3) implies that also $E_2 \setminus E_1 \in \mathcal{C}_F$, which implies that

$$|\mu|(E_1 \cup E_2) = |\mu|(E_1) + |\mu|(E_2 \setminus E_1)$$

$$\leq 2\alpha$$

$$< \infty,$$

and consequently, $|\mu|(F \setminus (E_1 \cup E_2)) = \infty$ (as $|\mu|(F) = \infty$). Thus indeed $E_1 \cup E_2 \in \mathcal{C}_F$ as asserted.

By definition of α, we can find a sequence $E_n \in \mathcal{C}_F$ such that $\alpha - \frac{1}{n} < |\mu|(E_n) \leq \alpha$, and by the last paragraph, we may assume, without loss of generality, that $E_n \subset E_{n+1}$. It follows that if we set $E = \cup_n E_n$, then $E \subset F$ and $|\mu|(E) = \lim_n |\mu|(E_n) = \alpha < \infty$, which implies that $|\mu|(F \setminus E) = \infty$ so that in fact $E \in \mathcal{C}_F$.

Now apply assertion (1) above - with $(F \setminus E)$ in place of what was called E in (1) to find a decomposition $(F \setminus E) = A \bigsqcup B$, with $|\mu|(A) > 0$ and $|\mu|(B) = \infty$.

This would imply that $E \bigsqcup A \in \mathcal{C}_F$ and hence that

$$\alpha \geq |\mu|(E \bigsqcup A) = |\mu|(E) + |\mu|(A) = \alpha + |\mu|(A) > \alpha \ ;$$

this contradiction completes the proof of (2).

(3)   If $F \in \mathcal{B}$ and if $|\mu|(F) = \infty$, then there exists a partition $F = C \bigsqcup D$, $C, D \in \mathcal{B}$ such that $|\mu(C)| \geq 1$ and $|\mu|(D) = \infty$.

Proof of (3): Indeed, in the notation of (3), we can find a set $E \in \mathcal{C}_F$ such that $|\mu|(E) > 2$; hence there exists a partition $E = \bigsqcup_n E_n$ and $\sum_n |\mu(E_n)| > 2$. If, as in Lemma 7.1.1, we let A (resp., B) denote the union of those E_n's for which $\mu(E_n)$ is positive (resp., negative), then we find that

$$\mu(A) + |\mu(B)| > 2.$$

The proof of (3) is completed by choosing $C \in \{A, B\}$ such that

$$|\mu(C)| = \max\{|\mu(A)|, |\mu(B)|\}$$

and setting $D = F \setminus C$.

Now the desired contradiction (that will complete the proof of the proposition) is obtained by the repeated application of assertion (3) above. Indeed, such a repeated application of (3) yields sequences $\{C_n\}_{n=1}^{\infty}, \{D_n\}_{n=0}^{\infty}$ of $\mathcal{B}$-measurable sets such that:

(a) $D_0 = \Omega$;
(b) $D_{n-1} = C_n \coprod D_n \ \forall \ n \geq 1$; and
(c) $|\mu(C_n)| \geq 1$ and $|\mu|(D_n) = \infty$.

Finally, we have a sequence $\{C_n\}_{n=1}^{\infty}$ of pairwise disjoint measurable sets such that $|\mu(C_n)| \geq 1 \ \forall \ n$. However the existence of such a sequence is ruled out by Lemma 7.1.1 and this contradiction completes the proof of the proposition. $\qquad \square$

Definition 7.1.5   *For a finite real measure μ, define $\mu_{\pm} = \frac{1}{2}(|\mu| \pm \mu)$.*

The inequality $|\mu|(E) \geq |\mu(E)|$ clearly implies that μ_+ and μ_- are finite positive measures and we have

$$|\mu| = \mu_+ + \mu_- \ \Rightarrow \ |\mu| - \mu_{\pm} = \mu_{\mp}. \tag{7.1.4}$$

Corollary 7.1.6   *Any finite complex measure μ can be decomposed into a linear combination of finite positive measures $\mu_i, 1 \leq i \leq 4$ as follows:*

$$\mu = (\mu_1 - \mu_2) + i(\mu_3 - \mu_4).$$

Further, when μ is real, we may obviously assume that $\mu_3 = \mu_4 = 0$.

Proof: First consider the case of real μ. In this case the desired decomposition is

$$\mu = \mu_+ - \mu_-.$$

For complex μ, apply the above decomposition to Re μ and Im μ. $\qquad \square$

A few remarks concerning 'finite complex measures' are now in order.

Remark 7.1.7   *Let μ be a finite complex measure. Even then we define the set-function $|\mu|$ by exactly the same formula as in the real case. The proof we have given of Lemma 7.1.3 can be repeated verbatim to show that the total variation* *of a finite complex measure is a positive measure. If μ_i, $1 \leq i \leq 4$ are as in Corollary 7.1.6, then it is easy to see that if $\Omega = \bigsqcup_n \Omega_n$ is any measurable partition, then,*

$$\sum_{n=1}^{\infty} |\mu(\Omega_n)| \leq \sum_{n=1}^{\infty} \sum_{i=1}^{4} \mu_i(\Omega_n)$$

$$= \sum_{i=1}^{4} \mu_i(\Omega)$$

so that on allowing the partition $\{\Omega_n\}_{n=1}^{\infty}$ to vary, we find that

$$|\mu|(\Omega) \leq \sum_{i} \mu_i(\Omega) < \infty,$$

i.e., the conclusion of Proposition 7.1.4 is also valid for finite complex measures.

7.2 L^p **Spaces**

This section is devoted to the definition and basic properties of the so-called L^p spaces. Throughout this section we shall assume that $(\Omega, \mathcal{B}, \mu)$ is a σ-finite (positive) measure space. We begin with a basic definition and proposition.

Definition 7.2.1 *Let $1 \leq p \leq \infty$.*

(1) *Define $\mathcal{L}^p(\Omega, \mathcal{B}, \mu)$ to be the class of complex-valued measurable functions $f : \Omega \to \mathbb{C}$ such that*

   (i) *$|f|^p \epsilon \mathcal{L}^1(\Omega, \mathcal{B}, \mu)$, i.e., $\int_{\Omega} |f|^p d\mu < \infty$ - if $1 \leq p < \infty$; and*
   (ii) *f is bounded μ-a.e. if $p = \infty$, i.e., if there exists a set $E \in \mathcal{B}$ (which will typically depend upon f) and a constant $M > 0$ such that $\mu(E) = 0$ and $|f(\omega)| \leq M$ whenever $\omega \notin E$.*

(2) *If $f \in \mathcal{L}^p(\Omega, \mathcal{B}, \mu)$, define*

$$\|f\|_p = \left(\int_{\Omega} |f|^p d\mu \right)^{\frac{1}{p}} \text{ if } 1 \leq p < \infty \tag{7.2.5}$$

and

$$\|f\|_{\infty} = \inf\{M > 0 : |f| \leq M \ \mu\text{-a.e.}\}, \text{ if } p = \infty . \tag{7.2.6}$$

Define $L^p(\Omega, \mathcal{B}, \mu)$ to be the set of equivalence classes in $\mathcal{L}^p(\Omega, \mathcal{B}, \mu)$, where, as usual, two functions are considered to be equivalent if they agree μ-a.e. It must be clear that $\| \cdot \|_p$ descends to a well-defined function on $L^p(\Omega, \mathcal{B}, \mu)$.

The first order of business is to prove that $\|\cdot\|_p$ is a norm on L^p for $1 \le p \le \infty$. The case of finite p is somewhat subtle and relies on a fundamental inequality.

Proposition 7.2.2   **(Hölder's inequality)** *Suppose that $\frac{1}{p} + \frac{1}{q} = 1$, where $1 < p < \infty$. Then, if $f \in L^p$ and $g \in L^q$, it follows that $fg \in L^1$, and we have:*

$$\left| \int_\Omega fg\,d\mu \right| \le \left(\int_\Omega |f|^p d\mu \right)^{\frac{1}{p}} \left(\int_\Omega |g|^q d\mu \right)^{\frac{1}{q}}. \tag{7.2.7}$$

Proof: To start with, note that the inequality is obvious if $\|f\|_p$ or $\|g\|_q$ is zero. So we assume the contrary.

Now, the logarithm is a 'concave function' on the positive line. (The chord joining two points on the graph $y = \log x$ always lies below the curve.) Hence, whenever $0 < t < 1$ and $a, b > 0$, we have:

$$t \log a + (1 - t) \log b \le \log(ta + (1 - t)b).$$

In other words,

$$a^t b^{1-t} \le ta + (1 - t)b. \tag{7.2.8}$$

(Since the logarithm is defined only for positive numbers, the above proof establishes this inequality only for strictly positive a, b; however, the inequality clearly extends to the case of non-negative a, b.)

Now apply the above inequality with $t = 1/p$, $a = \left(\frac{|f|}{\|f\|_p} \right)^p$ and $b = \left(\frac{|g|}{\|g\|_q} \right)^q$ to obtain

$$\frac{|f|}{\|f\|_p} \frac{|g|}{\|g\|_q} \le \frac{1}{p} \left(\frac{|f|}{\|f\|_p} \right)^p + \frac{1}{q} \left(\frac{|g|}{\|g\|_q} \right)^q.$$

Finally, integrate to obtain

$$\int \frac{|f|}{\|f\|_p} \frac{|g|}{\|g\|_q} d\mu \le \int \left(\frac{|f|^p}{p\|f\|_p^p} + \frac{|g|^q}{q\|g\|_q^q} \right) d\mu$$

$$= \frac{1}{p} + \frac{1}{q} = 1$$

as desired.   □

Remark 7.2.3

(i)   *Hölder's inequality can be compactly written thus:*

$$\|fg\|_1 \le \|f\|_p \|g\|_q,$$

where, of course, p and q are related as above. It is customary to call p and q conjugate indices if they are related as in Hölder's inequality.

(ii)   *It should be noted that the equation $\frac{1}{p} + \frac{1}{q} = 1$ is formally satisfied if we substitute $p = 1, q = \infty$, and that the inequality in (i) continues to hold for this specialisation of the indices. Consequently, one also sometimes thinks of ∞ as the conjugate of 1. In the rest of the chapter, if symbols p and q occur together, it will be assumed that they are conjugate indices.*

(iii) *It should also be noted that the 'infimum' in the definition of $\| \cdot \|_\infty$ is actually a minimum; this is because we can find μ-null sets E_n such that $sup_{\omega \notin E_n} |f(\omega)| < \|f\|_\infty + \frac{1}{n}$; but then $E = \cup_{n=1}^\infty E_n$ is a μ-null set with the property that $sup_{\omega \notin E} |f(\omega)| = \|f\|_\infty$.*

(iv)   *If $\|f_n - f\|_p \to 0$ with $p \in [1, \infty]$, then $\{f_n\}$ converges to f in measure. The proof is a verbatim copy of the proof of Proposition 5.2.2(b) with 'r' replaced everywhere by 'p'.*

We single out a useful consequence of Hölder's inequality as a corollary.

Corollary 7.2.4 *If $\mu(\Omega) < \infty$ and if $1 \leq s \leq r \leq \infty$, then $L^r(\Omega, \mathcal{B}, \mu) \subset L^s(\Omega, \mathcal{B}, \mu)$; in fact, we have*

$$\|\phi\|_s \leq (\mu(\Omega))^{\frac{1}{s} - \frac{1}{r}} \|\phi\|_r \ \forall \ \phi \ .$$

Proof: Suppose $\phi \in L^r$; set $f = |\phi|^s$, $p = \frac{r}{s}$, $g \equiv 1$ in Hölder's inequality. The assumption $s \leq r$ implies that $p \geq 1$, while the finiteness assumption ensures that the constant function g belongs to L^q for all $q \geq 1$. Hence we find that

$$\|\phi\|_s^s = \|fg\|_1 \leq \|f\|_p \|g\|_q = \|\phi\|_r^s \, (\mu(\Omega))^{\frac{r-s}{r}} \ ,$$

and the desired inequality is the sth root of the above inequality.   □

Proposition 7.2.5 *Let $1 \leq p \leq \infty$. Then ($\| \cdot \|_p$ is a norm and, in fact), $L^p(\Omega, \mathcal{B}, \mu)$ is a Banach space, i.e., it is a complete normed vector space.*

Proof: We first dispose of the easier case of $p = \infty$. Suppose $f, g \in \mathcal{L}^\infty(\Omega, \mathcal{B}, \mu)$ and $a \in \mathbb{C}$. By Remark 7.2.3 (iii), we may find null-sets F and G such that $|f(\omega)| \leq \|f\|_\infty$ if $\omega \notin F$, and $|g(\omega)| \leq \|g\|_\infty$ if $\omega \notin G$. Set $E = F \cup G$ and note that E is a null set such that if $\omega \notin E$, we have $|(af + g)(\omega)| \leq |a| \|f\|_\infty + \|g\|_\infty$; hence $(af + g) \in \mathcal{L}^\infty(\Omega, \mathcal{B}, \mu)$ and $\|af + g\|_\infty \leq |a| \|f\|_\infty + \|g\|_\infty$. It is easy to verify that $\| \cdot \|_\infty$ satisfies the other requirements for a norm on L^∞, viz.

$$\|f\| \geq 0$$

$$\|f\| = 0 \Leftrightarrow f = 0$$

$$\|af\| = |a| \, \|f\|.$$

(Note that $\| \cdot \|_\infty$ does not define a norm on $\mathcal{L}^\infty(\Omega, \mathcal{B}, \mu)$ since it is not positive definite.)

Since the union of two μ-null sets is again a μ-null set, and since the sum of two bounded functions is again bounded, it is clear that $\mathcal{L}^\infty(\Omega, \mathcal{B}, \mu)$ (and hence also $L^\infty(\Omega, \mathcal{B}, \mu)$) is a vector space. That this space is a normed space under $\| \cdot \|_\infty$ which is complete in that norm is a consequence of the (easily verified) fact that the collection of bounded functions on a set form a complete normed space with respect to the 'sup norm'.

Now fix $1 \le p < \infty$. Suppose $f, g \in \mathcal{L}^p(\Omega, \mathcal{B}, \mu)$. Note that $p = (p-1)q$, and hence

$$\int_\Omega |f + g|^p d\mu \le \int |f + g|^{p-1}(|f| + |g|)d\mu$$

$$\le \| |f + g|^{p-1} \|_q \, \| f \|_p + \| |f + g|^{p-1} \|_q \, \| g \|_p$$

$$= \left(\int_\Omega |f + g|^p d\mu \right)^{\frac{1}{q}} (\| f \|_p + \| g \|_p).$$

Thus, we find that

$$\left(\int |f + g|^p d\mu \right)^{1-\frac{1}{q}} \le \| f \|_p + \| g \|_p,$$

i.e., $\| f + g \|_p \le \| f \|_p + \| g \|_p$. (This argument works only for $f + g \ne 0$ but the desired inequality holds trivially in that case.) Thus, the triangle inequality is satisfied by $\| \cdot \|_p$. It is easy to verify that $\| \cdot \|_p$ satisfies the other requirements for a norm on L^p. The triangle inequality further shows that $\mathcal{L}^p(\Omega, \mathcal{B}, \mu)$ (and hence $L^p(\Omega, \mathcal{B}, \mu)$) is a vector space.

Finally, suppose $\{f_n\}$ is a Cauchy sequence in $\mathcal{L}^p$, i.e., the sequence satisfies $\lim_{n,m \to \infty} \| f_n - f_m \|_p = 0$. We may pick a subsequence $\{f_{n_k}\}_{k=1}^\infty$ with the property that

$$n, m \ge n_k \Rightarrow \| f_n - f_m \|_p < \frac{1}{2^k}.$$

In particular, we have $\| f_{n_{k+1}} - f_{n_k} \|_p < \frac{1}{2^k}$. Now consider the 'series' $f_{n_1} + \sum_{k=1}^\infty (f_{n_{k+1}} - f_{n_k})$ whose kth partial sum is just f_{n_k}.

Define $g_r = |f_{n_1}| + \sum_{k=1}^{r-1} |f_{n_{k+1}} - f_{n_k}|$ and note that $\| g_r \|_p \le \| f_{n_1} \|_p + \sum_{k=1}^{r-1} \| f_{n_{k+1}} - f_{n_k} \|_p$. Also, $\{g_r : r = 1, 2, 3, \dots\}$ is an increasing sequence of non-negative functions satisfying $|f_{n_k}| \le g_k \; \forall k$. Let $g = \sup_k g_k$ so that $|f_{n_k}| \le g \; \forall k$ as well. Finally, we have, thanks to monotone convergence,

$$\int g^p d\mu = \lim_{r \to \infty} \int g_r^p d\mu$$

$$= \lim_{r \to \infty} \|g_r\|_p^p$$

$$\leq \lim_{r \to \infty} \left(\|f_{n_1}\|_p + \sum_{k=1}^{r-1} \|f_{n_{k+1}} - f_{n_k}\|_p \right)^p$$

$$\leq (\|f_{n_1}\|_p + 1)^p$$

so that $g \in L^p$.

The finiteness of $\int g^p d\mu$ implies that g is finite a.e.; this shows that the sequence $\{f_{n_k}\}_{k=1}^{\infty}$ converges a.e. Define

$$f(\omega) = \begin{cases} \lim_k f_{n_k}(\omega) & \text{if this limit exists} \\ 0 & \text{otherwise} \end{cases}.$$

Then, $f_{n_k} \to f$ a.e. and since $|f_{n_k}| \leq g \in L^p$ a.e., we may deduce from Lebesgue's dominated convergence theorem that the sequence $\{f_{n_k}\}_{k=1}^{\infty}$ converges in $\| \cdot \|_p$ to f. It is easy to see that a Cauchy sequence converges to a limit f only if some subsequence converges to f; thus $\|f_n - f\|_p \to 0$, as desired.   □

Note that the above proof furnishes a sort of 'converse' to the dominated convergence theorem: *if a sequence converges in L^p, then it has a subsequence which is dominated pointwise (a.e.) by a function in L^p.*

We next exhibit a most natural dense subspace of $L^p(X, \mathcal{B}, \mu)$—at least when μ is a finite positive measure defined on the Borel σ-algebra of a compact metric space X.

Proposition 7.2.6 *Let μ be a finite positive measure on the Borel σ-algebra $\mathcal{B}_X$ of a compact metric space X. Then $C(X)$ is dense in $L^p(X)$ for $1 \leq p < \infty$.*

Proof: We may assume without loss of generality that $\mu(X) = 1$. Since simple functions are dense in $L^p(X)$, it is clearly enough to show that if $E \in \mathcal{B}_X$ and $\epsilon > 0$, then there exists a continuous function f such that $\|f - 1_E\|_p < \epsilon$. For this, begin by appealing to Proposition 1.7.2 to find an open set U and a compact set K such that $K \subset E \subset U$ and $\mu(U \setminus K) < \epsilon^p$; then, by Ex. 1.7.1(5) we can find a continuous function $f : X \to [0, 1]$ such that

$$f(x) = \begin{cases} 1 & \text{if } x \in K \\ 0 & \text{if } x \in X \setminus U. \end{cases}$$

Notice that $f(x) = 1_E(x)$ whenever $x \in K \cup (X \setminus U)$ and that $0 \le f^p$, $(1-f)^p \le 1$; deduce that

$$\int |f - 1_E|^p d\mu = \int_{U \setminus K} |f - 1_E|^p d\mu$$

$$= \int_{U \setminus E} f^p d\mu + \int_{E \setminus K} (1 - f)^p d\mu$$

$$\le \mu(U \setminus K)$$

$$< \epsilon^p$$

as desired. $\square$

We shall conclude this section with a few observations about what is probably the best understood (and arguably the most important) of the L^p spaces, namely, L^2. This is the only one whose norm 'comes from an underlying inner-product'. If $f, g \in L^2$, define

$$\langle f, g \rangle = \int f \bar{g} d\mu. \tag{7.2.9}$$

(Notice that the index $p = 2$ is its own conjugate index, and Hölder's inequality yields, in this case, the so-called **Cauchy–Schwarz inequality**

$$|\langle f, g \rangle| \le \|f\|_2 \|g\|_2, \tag{7.2.10}$$

which guarantees that the left side of (7.2.9) is indeed a finite complex number called the *inner product* of f and g, which satisfies, for all $f, g, h \in L^2$ and $a, b \in \mathbb{C}$:

- $\langle f, f \rangle = \|f\|_2^2$
- $\langle af + bh, g \rangle = a\langle f, g \rangle + b\langle h, g \rangle$
- $\langle f, g \rangle = \overline{\langle g, f \rangle}$.

In fact, L^2 is the prototypical **Hilbert space**, i.e., complete normed space whose norm 'comes from an inner-product' as above.

In order for our treatment in the next section to be complete, we will need one general fact about Hilbert spaces the proof of which can be found in standard texts of functional analysis; we state it here after making the necessary definitions.

Definition 7.2.7   *Suppose X is a normed vector space (such as L^p, for instance) over the field of complex numbers. A mapping $\phi : X \to \mathbb{C}$ is called a* **bounded linear functional** *if it satisfies two conditions, $\forall x, y \in X, a, b \in \mathbb{C}$:*

- *(linearity) $\phi(ax + by) = a\phi(x) + b\phi(y)$; and*

- *(boundedness)* $|\phi(x)| \leq K \|x\|$ *for some constant* $K > 0$.

The collection of all such bounded linear functionals on X is usually denoted by X^*.

The norm of such a bounded linear functional is defined by

$$\|\phi\| = \inf\{K > 0 : |\phi(x)| \leq K\|x\| \; \forall x\}.$$

It is a fact that X^* is a vector space and that the above definition yields a norm with respect to which X^* is complete. The Banach space X^* is called the **dual space** of X.

Remark 7.2.8 *If* $X = L^p(\Omega, \mathcal{B}, \mu)$, *it follows from Hölder's inequality that each element* $g \in L^q$ *determines a bounded linear functional* ϕ_g *on* L^p *by the rule* $\phi_g(f) = \int fg d\mu$. *It is a fact (see Theorem 7.3.11) that if* $1 < p < \infty$, *every bounded linear functional on* L^p *arises in this manner. This fact will be proved using the Radon–Nikodym theorem, but the proof of the latter theorem will be based on the validity of the special case of the same fact for* $p = 2$; *and in order to avoid circular reasoning, we shall formulate the special case as a general fact (called the Riesz lemma) about Hilbert spaces, and direct the reader to standard texts (such as [Sim] or [Sun], for instance) for a proof.*

Theorem 7.2.9   **(Riesz lemma)** *If* ϕ *is a bounded linear functional on a Hilbert space* $\mathcal{H}$, *then there exists a unique vector* $y \in \mathcal{H}$ *such that* $\phi(x) = \langle x, y \rangle \; \forall x$. *Further,* $\|y\| = \|\phi\|$.

Actually, the form in which we shall be using this result is this: if ϕ is a bounded conjugate-linear functional on L^2—meaning that $\phi : L^2 \rightarrow \mathbb{C}$ is such that $f \mapsto \overline{\phi(f)}$ is a bounded linear functional—then there exists a unique $g \in L^2$ such that $\phi(f) = \int g\bar{f}d\mu$.

7.3   Radon–Nikodym Theorem

This section will be devoted to the fundamental **Radon–Nikodym theorem** and some of its consequences. We begin with a simple observation.

Proposition 7.3.1   *Let* $(\Omega, \mathcal{B}, \mu)$ *be a* σ-*finite measure space, and let* $f : \Omega \rightarrow [0, \infty)$ *be a non-negative function which is finite* μ-*a.e. Then the equation*

$$\nu(E) = \int_E f d\mu = \int 1_E f d\mu \tag{7.3.11}$$

defines a σ-*finite measure on* $(\Omega, \mathcal{B})$, *which is sometimes called the* '**indefinite integral**' *of* f; *this measure has the property that*

$$E \in \mathcal{B}, \ \mu(E) = 0 \ \Rightarrow \ \nu(E) = 0 \ . \tag{7.3.12}$$

Proof: Let $\Omega = \cup_n \Omega_n$ be a decomposition of Ω as an increasing sequence of $\mathcal{B}$-measurable sets such that $\mu(\Omega_n) < \infty \ \forall n$, and define $E_n = \{\omega \in \Omega_n : f(\omega) \le n\}$. It is then easy to see that ν is a positive measure on $(X, \mathcal{B})$ with the property that $\nu(E_n) < \infty$, as also that $\{E_n\}_{n=1}^{\infty}$ is an increasing sequence of sets in $\mathcal{B}$ such that $\Omega = \cup_n E_n$. Thus ν is indeed a σ-finite measure. Finally, it is obvious that $\mu(E) = 0 \ \Rightarrow \ \nu(E) = 0$.

If positive measures μ and ν are related as in the condition (7.3.12), the measure ν is said to be **absolutely continuous** with respect to μ.

The content of the Radon–Nikodym theorem is that, in the presence of σ-finiteness, all absolutely continuous measures arise as indefinite integrals.

Theorem 7.3.2   (Radon–Nikodym Theorem) *Suppose μ and ν are σ-finite measures defined on the measurable space $(\Omega, \mathcal{B})$. Suppose ν is absolutely continuous with respect to μ. Then there exists a non-negative function g with the property that*

$$\nu(E) = \int_E g \, d\mu \ \forall \ E \in \mathcal{B};$$

the function g is uniquely determined μ-a.e. by the above requirement.

Further, it is true that if f is any non-negative measurable function on X, then

$$\int f \, d\nu = \int f g \, d\mu.$$

Proof: Define $\rho = \mu + \nu$.

Case (i): $\rho(\Omega) < \infty$ (or equivalently, both μ and ν are finite measures).

In this case, consider the map $\phi : L^2(\Omega, \mathcal{B}, \rho) \to \mathbb{C}$ defined by

$$\phi(f) = \int_{\Omega} \bar{f} \, d\nu \tag{7.3.13}$$

and note that, under the finiteness assumption, we may deduce from the Cauchy–Schwarz inequality that

$$|\phi(f)| = \left| \int \bar{f} 1 \, d\nu \right| \le \|\bar{f}\|_{L^2(\nu)} \|1\|_{L^2(\nu)}$$

$$= (\nu(\Omega))^{\frac{1}{2}} \|f\|_{L^2(\nu)} \le (\nu(\Omega))^{\frac{1}{2}} \|f\|_{L^2(\rho)}$$

since $\nu \le \rho$. Thus ϕ is a bounded (clearly) conjugate-linear functional on $L^2(\rho)$; so by (the conjugate-linear version of) Riesz' lemma, we may conclude that there exists an element $h \in L^2(\rho)$ such that

$$\int \bar{f} dv = \phi(f) = \int h\bar{f} d\rho \ \forall f \in L^2(\rho).$$
(7.3.14)

In particular, since bounded functions are integrable with respect to any finite measure, we see that if f is any non-negative bounded measurable function on X, then

$$\int f dv = \int h f d\rho = \int h f d\mu + \int h f dv,$$

i.e.,

$$\int f(1-h)dv = \int f h d\mu$$
(7.3.15)

for every non-negative bounded measurable function f.

Observe that (7.3.14) (applied with 1_E in place of f) implies that $v(E) = \int_E h d\rho \geq 0 \ \forall E \in \mathcal{B}$, and hence we conclude that $h \geq 0$ ρ-a.e. Hence also $h \geq 0$ μ-a.e.

Note also that if we apply (7.3.15) to $f = 1_M$, where $M = \{h \geq 1\}$, we find that

$$0 \leq \mu(M) \leq \int_M h d\mu = \int_M (1-h)dv \leq 0,$$

which implies that $\mu(M) = 0$. Hence we see that $0 \leq h < 1$ μ-a.e. (and hence also v-a.e.)

This allows us to conclude that (7.3.15) is valid for any non-negative measurable function f (by virtue of its validity for each function $f 1_{\{f \leq n\}}$, and (two applications of) the monotone convergence theorem).

Now, if k is an arbitrary non-negative measurable function, apply (7.3.15) with $f = \frac{k}{1-h}$, which makes sense a.e. to obtain

$$\int k dv = \int f(1-h)dv$$

$$= \int f h d\mu$$

$$= \int k \frac{h}{1-h} d\mu.$$

In other words, if we define $g = \frac{h}{1-h}$, we find that $g \geq 0$ μ-a.e. and that

$$\int k dv = \int kg d\mu$$
(7.3.16)

for every non-negative measurable function k. In particular, setting $k = 1_E$ yields

$$v(E) = \int_E g \, d\mu.$$

For uniqueness, appeal to Ex. 2.3.6. This completes the proof of the theorem in this case.

Case (ii): ρ arbitrary.

Let $\Omega = \cup_n E_n = \cup_n F_n$ be decompositions of Ω as increasing unions with the property $\mu(E_n), \nu(F_n) < \infty \ \forall n$. Let $\Omega_n = E_n \cap F_n$. Then $\Omega = \cup \Omega_n$ is a decomposition of Ω as an increasing union with the property $\rho(\Omega_n) < \infty \ \forall n$. The validity of the theorem in this case follows from the validity of the theorem for the restriction of the measures to Ω_n (which follows from the already proved Case (i) and the uniqueness assertion in that case). □

Remark 7.3.3

(a) *The function that was denoted by g in the foregoing theorem is called the* **Radon–Nikodym derivative of** v **w.r.t.** μ *and usually denoted by the symbol $\frac{dv}{d\mu}$. The reason for the terminology will be clarified in the next section; the reason for the notation becomes clear upon re-writing the last identity in the statement of the theorem in the new notation. Thus,*

$$\int f \, dv = \int f \frac{dv}{d\mu} \, d\mu. \qquad (7.3.17)$$

(b) *If μ, ν are as in the theorem and if λ is another σ-finite measure such that μ is absolutely continuous with respect to λ, then it is easily seen that ν is absolutely continuous with respect to λ. Further, the uniqueness assertion about the Radon–Nikodym derivative, coupled with (7.3.17), shows that the following* **chain rule for derivatives** *is valid:*

$$\frac{dv}{d\lambda} = \frac{dv}{d\mu} \frac{d\mu}{d\lambda}.$$

(c) *It follows from (b) that if ν is absolutely continuous with respect to μ, and μ is absolutely continuous with respect to ν, then we must have*

$$\frac{dv}{d\mu} = \left(\frac{d\mu}{dv} \right)^{-1},$$

and in particular, $\frac{dv}{d\mu} > 0$ μ-a.e. It is left as an exercise to the reader to show that conversely, if ν is absolutely continuous with respect to μ, and if $\frac{dv}{d\mu} > 0 \mu$-a.e., then it must be the case that μ also is absolutely continuous with respect to ν. When this happens, it is customary to say that μ and ν are **mutually absolutely continuous.**

Ex. 7.3.4

(a) Let $\mu_\pm$ denote the 'upper' and 'lower' variation measures defined in Definition 7.1.5. Then show that μ_- and μ_- are both absolutely continuous with respect to $|\mu|$. Conclude that there exists an essentially unique (use Ex. 2.3.6) $g \in L^1(|\mu|)$, denoted by $\frac{d\mu}{d|\mu|}$, such that

$$\mu(E) = \int_E g\,d|\mu| \ \forall E \in \mathcal{B}. \tag{7.3.18}$$

(b) With g as above, define ρ to be the measure such that $\frac{d\rho}{d|\mu|} = |g|$.

  (i) Use the definition of $|\mu|$ to show that $|\mu|(E) \leq \rho(E)$ for all E; conclude that ρ and $|\mu|$ are mutually absolutely continuous and that $|g| \geq 1$ $|\mu|$-a.e.

  (ii) Conversely, use the fact that $|\mu(E)| \leq |\mu|(E)$ for all E to conclude that $|g| = 1$ $|\mu|$-a.e.

(c) Set $A_\pm = g^{-1}(\{\pm 1\})$. Show that $\frac{d\mu_-}{d|\mu|} = \pm 1_{A_\pm}$.

The measures $\mu_\pm$ exhibit the property of 'mutual singularity' as described in the next definition.

Definition 7.3.5 *Two positive measures $\mu_i, i = 1, 2$, defined on the same measurable space $(\Omega, \mathcal{B})$, are said to be* **mutually singular** *if there exists a measurable partition $\Omega = \Omega_1 \bigsqcup \Omega_2$ such that μ_i is 'supported on' Ω_i in the sense that $\mu_i(\Omega \setminus \Omega_i) = 0$.*

Ex. 7.3.6 If μ is a finite real measure and if μ_1 and μ_2 are mutually singular finite positive measures such that $\mu = \mu_1 - \mu_2$, then show that $\mu_1 = \mu_+$ and $\mu_2 = \mu_-$, where $\mu_\pm$ are as in Ex. 7.3.4.

All the pieces are now in place for the following result.

Proposition 7.3.7 (**Hahn decomposition**) *Any finite real measure μ admits a unique decomposition $\mu = \mu_+ - \mu_-$ as a difference of two mutually singular finite positive measures $\mu_\pm$; further, $\mu_\pm$ are the 'upper' and 'lower' variation measures defined in Definition 7.1.5.*

Corollary 7.3.8 *Any finite complex measure μ admits the canonical decomposition*

$$\mu = (\text{Re } \mu)_+ - (\text{Re } \mu)_- + i((\text{Im } \mu)_+ - (\text{Im } \mu)_-)$$

as a linear combination of finite positive measures.

Ex. 7.3.9   Use Corollary 7.3.8 to define the notion of integration with respect to any finite complex measure and prove a version of the Radon–Nikodym theorem for complex measures. (*Hint:* If μ and ν are finite complex measures, define ν to be absolutely continuous with respect to μ if it is the case that $|\nu|$ is absolutely continuous with respect to $|\mu|$.)

We next discuss a useful consequence of the Radon–Nikodym theorem. (We state and prove the theorem for finite positive measures, and leave the formulation as well as proof for σ-finite or finite complex measures to the readers.)

Theorem 7.3.10   (**The Lebesgue–Nikodym theorem**) *Let μ and ν be finite positive measures defined on $(\Omega, \mathcal{B})$.*

(1)   *Then there exists a decomposition $\nu = \nu_1 + \nu_2$ into finite positive measures such that (i) ν_1 is absolutely continuous with respect to μ, and (ii) ν_2 and μ are mutually singular.*

(2)   *The above decomposition is unique; if $\nu = \nu_1' + \nu_2'$ is another such decomposition, then $\nu_i = \nu_i'$.*

Proof:

(a)   Let $\rho = \mu + \nu$ and let $g = \frac{d\nu}{d\rho}$. Notice first of all that $0 \leq g \leq 1 \rho$-a.e. Define $\Omega_2 = g^{-1}(\{1\})$, $\Omega_1 = \Omega \setminus \Omega_2$ and let ν_i be the measure given by $\frac{d\nu_i}{d\rho} = g 1_{\Omega_i}$. It is immediately clear that $\nu = \nu_1 + \nu_2$. Observe, next, that

$$\nu_2(\Omega_2) = \int_{\Omega_2} g d\rho = \rho(\Omega_2).$$

Since $\rho = \mu + \nu_1 + \nu_2$, deduce that $\mu(\Omega_2) = \nu_1(\Omega_2) = 0$. Since, clearly, $\nu_2(\Omega_1) = 0$, we see that indeed μ and ν_2 are mutually singular. Finally, observe that if $E \in \mathcal{B}$,

$$\mu(E) = 0 \Rightarrow \mu(F) = 0 \; \forall F \subset E, F \in \mathcal{B}$$

$$\Rightarrow \nu(F) = \rho(F) \; \forall F \subset E, F \in \mathcal{B}$$

$$\Rightarrow \int_F g d\rho = \rho(F) \; \forall F \subset E, F \in \mathcal{B}$$

$$\Rightarrow g 1_E = 1 \rho - a.e.$$

$$\Rightarrow \rho(E \cap \Omega_1) = 0$$

$$\Rightarrow \nu_1(E) = 0,$$

and hence indeed ν_1 is absolutely continuous with respect to μ.

(b)   Suppose $\nu = \nu'_1 + \nu'_2$ is another such decomposition. Then there exists another decomposition $\Omega = \Omega'_1 \bigsqcup \Omega'_2$ such that μ is also supported on Ω'_1, and ν'_2 is supported on Ω'_2. It follows that if we let $S = \Omega_1 \cap \Omega'_1$, then μ is actually supported on S and that ν_2 and ν'_2 are supported on $\Omega \setminus S$. On the other hand, ν_1 and ν'_1 being absolutely continuous with respect to μ, they have to be supported on S; so this must also be true of $|\nu'_2 - \nu_2| = |\nu_1 - \nu'_1|$; but $|\nu'_2 - \nu_2| \leq \nu_2 + \nu'_2$, and so $|\nu'_2 - \nu_2|$ must also be supported on $\Omega \setminus S$. Since only the zero measure can be supported on a set as well as on its complement, the proof is complete.   □

Now we shall establish—as promised in Remark (7.2.8)—the L^p-analogue of the Riesz lemma.

Theorem 7.3.11   *Let* $(\Omega, \mathcal{B}, \mu)$ *be a σ-finite measure space. Let* $1 \leq p < \infty$ *and let q be the conjugate index of p.*

(1)   *If $g \in L^q(\Omega, \mathcal{B}, \mu)$, then the equation*

$$\phi_g(f) = \int fg \, d\mu \qquad (7.3.19)$$

defines a bounded linear functional ϕ_g on $L^p(\Omega, \mathcal{B}, \mu)$.

(2)   *Conversely, if $\phi \in (L^p(\Omega, \mathcal{B}, \mu))^*$, then there exists a unique $g \in L^q(\Omega, \mathcal{B}, \mu)$ such that $\phi = \phi_g$.*

(3)   *Further* $\|\phi_g\|_{(L^p(\Omega, \mathcal{B}, \mu))^*} = \|g\|_{L^q(\Omega, \mathcal{B}, \mu)}$. *(Thus, the mapping $g \mapsto \phi_g$ defines an 'isometric isomorphism' of the Banach space $L^q(\Omega, \mathcal{B}, \mu)$ onto the Banach space $(L^p(\Omega, \mathcal{B}, \mu))^*$.)*

Proof:

(1)   The case $p = 1$ is clear and requires no proof. So assume that $p > 1$. Hölder's inequality guarantees that ϕ_g is indeed a bounded linear functional on L^p.

(2)   We shall discuss the proof in the case when μ is a finite measure, and when $p > 1$. (The other cases, which only need minor modifications, are left to the reader.) Then,

$$E \in \mathcal{B} \Rightarrow 1_E \in L^p(\mu),$$

and hence the equation

$$\nu(E) = \phi(1_E)$$

is seen to meaningfully define a function $\nu : \mathcal{B} \to \mathcal{C}$. To see that ν is a finite complex measure on $\mathcal{B}$, we only need to check countable additivity; finite additivity follows from the linearity of ϕ; as for continuity from below, if $\{E_n\}$ is a sequence of sets which increases to a set E, then $\mu(E_n) \uparrow \mu(E)$ and so

$$\|\phi(1_{E_n}) - \phi(1_E)\| \le \|\phi\| \ \|1_{E_n} - 1_E\|_p = \|\phi\| \ (\mu(E) - \mu(E_n))^{\frac{1}{p}} \downarrow 0.$$

Further,

$$E \in \mathcal{B}, \mu(E) = 0 \Rightarrow 1_E = 0 \in L^p \Rightarrow \nu(E) = \phi(1_E) = 0.$$

Since this is valid for every measurable subset of E, this shows that $|\nu|$ (and hence ν) is absolutely continuous with respect to μ. Let $g = \frac{d\nu}{d\mu}$. Notice that if $f = \sum_{i=1}^n c_i 1_{E_i}$ is any simple function, then

$$\phi(f) = \sum_{i=1}^n c_i \phi(1_{E_i})$$

$$= \sum_{i=1}^n c_i \int 1_{E_i} d\nu$$

$$= \int f d\nu$$

$$= \int fg \, d\mu. \tag{7.3.20}$$

We wish to observe next that (7.3.20) continues to be valid for all bounded measurable functions f.

(For such an f, we can construct (by Remark 2.1.5) a sequence s_n of simple functions, such that $\|s_n - f\|_\infty \to 0$. Then also $\|s_n - f\|_p \to 0$. Hence

$$\phi(f) = \lim_{n\to\infty} \phi(s_n)$$

$$= \lim_{n\to\infty} \int s_n g d\mu$$

$$= \int fg d\mu \quad \text{by dominated convergence theorem.)} \tag{7.3.21}$$

Choose a sequence $\{h_n\}$ of non-negative simple functions which increases pointwise to $|g|$. Write $f_n = \frac{\bar{g}}{|g|} h_n^\alpha$ where $\alpha = \frac{1}{p-1}$, and argue that $\|f_n\|_p = (\|h_n\|_q)^{\frac{q}{p}}$, and that

$$\phi(f_n) = \int f_n g d\mu$$

$$= \int |g| h_n^\alpha d\mu$$

$$\geq \int h_n^{\alpha+1} d\mu$$

$$= \|h_n\|_q^q.$$

We thus see that

$$\|h_n\|_q^q \leq |\phi(f_n)| \leq \|\phi\| \|f_n\|_p = \|\phi\| (\|h_n\|_q)^{\frac{q}{p}}$$

and hence that $\|\phi\| \geq \|h_n\|_q$. We may finally conclude that

$$\|\phi\| \geq \sup_n \|h_n\|_q = \|g\|_q \tag{7.3.22}$$

so that $g \in L^q$. Then $\phi_g \in (L^p(\Omega, \mathcal{B}, \mu))^*$, which agrees with ϕ on the dense subspace of $L^p(\Omega, \mathcal{B}, \mu)$ consisting of bounded measurable functions. Therefore $\phi = \phi_g$.

(3)   Hölder's inequality in fact shows that $\|\phi_g\| \leq \|g\|_q$. To prove the reverse inequality we may as well assume $g \neq 0$ a.e. Consider the function defined by $f_0 = \frac{\bar{g}}{|g|} |g|^\alpha$, where $\alpha = \frac{1}{p-1}$. We have

$$\|f_0\|_p = (\|g\|_q)^{\frac{q}{p}}, \text{ and } \phi_g(f_0)(= \|g\|_q^q) = \|f_0\|_p \|g\|_q;$$

and these identities reveal that we must have $\|\phi_g\| \geq \|g\|_q$, thus establishing that $\|\phi_g\| = \|g\|_q$.   □

7.4   Change of Variables

This section will be devoted to a brief discussion of the basic *change of variable formula*. A reader without sufficient background in 'several variable calculus' can safely omit this section without it affecting her/his understanding of the subsequent sections if so desired; but should try and 'get a feel' for what this fundamental result says. In any case, nothing is really proved here, only a flavour of this result is attempted to be conveyed.

We start by addressing an abstract change of variable formula. Begin by recalling that if $(\Omega, \mathcal{B}, \mu)$ and $(\Omega', \mathcal{B}')$ is are measure spaces, and if $T : \Omega \to \Omega'$ is a measurable map, then $\mu \circ T^{-1}$ is the measure defined on $\mathcal{B}'$ by the rule

$$(\mu \circ T^{-1})(E') = \mu(T^{-1}(E')) .$$

We then have the following easy proposition whose proof we omit since it is an immediate consequence of Proposition 2.1.4 and the observation that the statement holds when $f = 1_{E'}$ by definition.

Proposition 7.4.1   *With the above notation, if* $f : \Omega' \to [0, \infty)$ *is a non-negative measurable function, then*

$$\int_{\Omega'} f d(\mu \circ T^{-1}) = \int_{\Omega} (f \circ T) d\mu.$$

Remark 7.4.2   *Suppose* $A : \mathbb{R}^n \to \mathbb{R}^n$ *is a linear mapping. Almost by definition of the determinant, we have*

$$|\det A| = \lambda(A(C)),$$

where C *is the 'unit cube' in* $\mathbb{R}^n$ *given by* $C = \{\sum_{i=1}^n \theta_i \mathbf{e_i} : 0 \le \theta_i \le 1\}$ *(where of course* $\mathbf{e_i}$ *is the* i*th standard basis vector with 1 in the* i*th coordinate and all others equal to zero). It follows that if* $B = rC + \mathbf{x_0}$ *is an arbitrary cube in* $\mathbb{R}^n$ *(with sides parallel to the axes), then we see from the translation invariance of* λ *that*

$$\lambda(A(B)) = \lambda(A(rC) + A(\mathbf{x_0})) = \lambda(A(r(C)) = \lambda(rA(C))$$

$$= r^n |\det A| = \lambda(B)|\det A|.$$

Thus, we see that $(\lambda \circ A)(E) = |\det A|\lambda(E)$ *for any cube* E *with sides parallel to the axes. It follows that the above assertion continues to hold for every open set* E *and hence also for every Borel set. (Why?) Deduce now from Proposition 7.4.1 that if* A *is an invertible linear transformation, then*

$$|\det A|^{-1} \int f(\mathbf{y}) d\lambda(\mathbf{y}) = \int f(A\mathbf{x}) d\lambda(\mathbf{x}).$$

What is usually called the change of variable theorem (or formula) is the analogue of the above equation for nonlinear maps. To formulate the result, we need to recall some facts concerning derivatives of functions of several variables.

So, suppose $f : U \to \mathbb{R}^m$ is a map where $U \subset \mathbb{R}^n$. Then there exists m functions $f_1, \ldots f_m : U \to \mathbb{R}$ such that $f(\mathbf{x}) = (f_1(\mathbf{x}), f_2(\mathbf{x}), \ldots, f_m(\mathbf{x})) \, \forall \mathbf{x} \in U$. Recall that if U is an open set, we say that f is differentiable at a point $\mathbf{x_0} \in U$ if there exists a linear map $f'(\mathbf{x_0}) : \mathbb{R}^n \to \mathbb{R}^m$ such that

$$\lim_{\|\mathbf{h}\| \to 0} \frac{\|f(\mathbf{x_0} + \mathbf{h}) - f(\mathbf{x_0}) - f'(\mathbf{x_0})\mathbf{h}\|}{\|\mathbf{h}\|} = 0;$$

in other words the linear map $f'(\mathbf{x_0})$ is the *best linear approximation at* $\mathbf{x_0}$ *to the map* f. If f is so differentiable at $\mathbf{x_0}$, then it is true that this linear map is described by the matrix $((\frac{\partial f_i}{\partial x_j}(\mathbf{x_0})))$ in the sense that

$$(f'(\mathbf{x_0})\mathbf{h})_i = \sum_{j=1}^{n} \frac{\partial f_i}{\partial x_j}(\mathbf{x_0})h_j.$$

As with the case of one-variable, we have the chain rule, i.e., if also $g : V \to \mathbb{R}^k$ is a map of an open neighbourhood of $f(\mathbf{x_0})$ into $\mathbb{R}^k$ which is differentiable at $f(\mathbf{x_0})$, then the composite function $g \circ f$ is differentiable at $\mathbf{x_0}$ and $(g \circ f)'(\mathbf{x_0}) = g'(f(\mathbf{x_0})) \cdot f'(\mathbf{x_0})$.

It easily follows from the definitions that if $A : \mathbb{R}^n \to \mathbb{R}^m$ is a linear map, then A is differentiable everywhere and that $A'(\mathbf{x_0}) = A$ for all $\mathbf{x_0}$.

Now, suppose $T : U \to V$ is a 1-1 map of an open subset U of $\mathbb{R}^n$ onto an open subset V of $\mathbb{R}^m$ and suppose that both T and the inverse map T^{-1} are differentiable. Then it would follow that for every $\mathbf{x} \in \mathbb{R}^n$ the linear maps $T'(\mathbf{x}) : \mathbb{R}^n \to \mathbb{R}^m$ and $(T^{-1})'(T(\mathbf{x})) : \mathbb{R}^m \to \mathbb{R}^n$ would have to be inverses of one another. In particular, we must have $m = n$, and

$$J_T(\mathbf{x}) = (J_{T^{-1}}(T(\mathbf{x})))^{-1}, \tag{7.4.23}$$

where we write $J_T(\mathbf{x}) = \det(T'(\mathbf{x}))$.

The crucial ingredient of the 'change of variable theorem' is the assertion that if

(i) T and T^{-1} are as in the preceding paragraph, and
(ii)   the map $\mathbf{x} \to T'(\mathbf{x})$ is continuous[1],

then the measures $\lambda \circ T$ and λ (on V) are mutually absolutely continuous and

$$\frac{d(\lambda \circ T^{-1})}{d\lambda}(\mathbf{y}) = |J_{T^{-1}}(\mathbf{y})| \ \forall \ y \in V.$$

Consequently, with $\mathbf{y} = T(\mathbf{x})$, it would follow that

$$\int_V f(\mathbf{y})d\lambda(\mathbf{y}) = \int_V \frac{f(\mathbf{y})}{|J_{T^{-1}}(\mathbf{y})|}d(\lambda \circ T^{-1})(\mathbf{y})$$

$$= \int_V \frac{f}{|J_{T^{-1}}|}d(\lambda \circ T^{-1})$$

---

[1]This is paraphrased by the statement that T is a C^1-diffeomorphism.

$$= \int_U \left(\frac{f}{|J_{T^{-1}}|} \circ T \right) d\lambda \quad \text{by Proposition 7.4.1}$$

$$= \int_U f(T\mathbf{x}) |J_T(\mathbf{x})| d\lambda(\mathbf{x}) \quad \text{by (7.4.23),}$$

which is the customary formulation of the change of variable theorem.

7.5   Differentiation

Definition 7.5.1   *Let $A \subset \mathbb{R}$. A collection of intervals $\mathcal{I}$ is said to be a **Vitali cover** of A if $\forall \epsilon > 0, \forall x \in A$ there is an $I \in \mathcal{I}$ such that $x \in I$ and $\lambda(I) < \epsilon$.*

Lemma 7.1   *Let A be set of finite Lebesgue measure and $\mathcal{I}$ be a Vitali cover of A. Then, given $\epsilon > 0$, there is a finite disjoint collection $\{I_1, I_2, \ldots I_n\}$ of intervals in $\mathcal{I}$ such that*

$$\lambda(A - \cup_{k=1}^n I_k) < \epsilon.$$

Proof: Without loss of generality we may assume that

(1)   any interval in the collection $\mathcal{I}$ is closed (this is possible because singletons have Lebesgue measure 0) and

(2)   there is an open set U of finite measure such that it contains A and every $I \in \mathcal{I}$.

We shall proceed by induction. Let I_1 be any interval in $\mathcal{I}$. Suppose $I_2, I_3, \ldots, I_n$ are already chosen. If $A \subset \cup_{k=1}^n I_k$, then we are done. Otherwise, let

$$s_n = \sup\{\lambda(I) | I \in \mathcal{I}, I \cap I_k = \emptyset, k = 1, 2, \ldots n\}.$$

By assumption (b), we know that $s_n < \infty$. Choose $I_{n+1} \in \mathcal{I}$, such that it is

(a)   disjoint from $I_1, I_2, \ldots, I_n$, and

(b) $\lambda(I_{n+1}) > \frac{s_n}{2}$.

There are two possibilities: (i) $\exists N$ such that $A \subset \cup_{k=1}^N I_k$, in which case the process stops. (ii) This inductive process generates an infinite sequence $\{I_n : n \in \mathbb{N}\}$ of intervals satisfying the above properties.

In case (i) we are done. So assume we are in case (ii). Observe that

$$\sum_{n=1}^\infty \lambda(I_n) = \lambda(\cup_{n=1}^\infty I_n) \leq \lambda(U) < \infty.$$

Let $\epsilon > 0$ be given. There exists a N such that $\sum_{n=N+1}^\infty \lambda(I_n) < \epsilon$. Consider the set $L = A - \cup_{n=1}^N I_n$. Since $\cup_{n=1}^N I_n$ is a closed set, $\mathcal{I}$ being a Vitali cover for

A, it follows that for any $x \in L$ there is an $I \in \mathcal{I}$ that contains x and does not intersect $\cup_{n=1}^{N} I_n$. Observe that

- if $I \cap I_k = \emptyset$ for all $k \leq n$ then $\lambda(I) \leq s_n < 2\lambda(I_{n+1})$
- $\lambda(I_n) \to 0$ as $n \to \infty$.

So there exists an integer k (necessarily greater than N) such that I intersects I_k; without loss of generality we may assume that k is the smallest such integer. Let m_k be the midpoint of I_k.

$$|x - m_k| \leq \lambda(I) + \frac{1}{2}\lambda(I_k) < s_{k-1} + \frac{1}{2}\lambda(I_k) \leq 2\lambda(I_k) + \frac{1}{2}\lambda(I_k)$$

$$= \frac{5}{2}\lambda(I_k).$$

Let J_k be an interval in $\mathbb{R}$ with midpoint m_k but having length five times the length of I_k. We have shown above that $L \subset \cup_{k=N+1}^{\infty} J_k$. Thus,

$$\lambda(L) \leq \lambda(\cup_{k=N+1}^{\infty} J_k) \leq \sum_{k=N+1}^{\infty} \lambda(J_k) = \sum_{k=N+1}^{\infty} 5\lambda(I_k) \leq 5\epsilon.$$

As ϵ was arbitrary, we are done.   □

Definition 7.5.2   *Let $a, b \in \mathbb{R}$. $f : [a, b] \to \mathbb{R}$ is said to be **differentiable** at a point $x \in [a, b]$ if*

$$f'(x) = \lim_{h \to 0} \frac{f(x + h) - f(x)}{h}$$

exists. We say a function is differentiable on $\mathbb{R}$ if it is differentiable at all $x \in \mathbb{R}$ and we call f' the derivative of f.

In the above definition, if $x = a$, we assume that $h \downarrow 0$ and if $x = b$ we assume that $h \uparrow 0$. Some of the other common variations used are

$$RDf(x) = \lim_{h \downarrow 0} \frac{f(x + h) - f(x)}{h},$$

$$D^+ f(x) = \limsup_{h \downarrow 0} \frac{f(x + h) - f(x)}{h},$$

$$D_+ f(x) = \liminf_{h \downarrow 0} \frac{f(x + h) - f(x)}{h},$$

$$LDf(x) = \lim_{h \uparrow 0} \frac{f(x+h) - f(x)}{h},$$

$$D^- f(x) = \limsup_{h \uparrow 0} \frac{f(x+h) - f(x)}{h},$$

$$D_- f(x) = \liminf_{h \uparrow 0} \frac{f(x+h) - f(x)}{h}.$$

Clearly, $RDf(x)$ exists if and only if $D^+ f(x)$ and $D_+ f(x)$ exist and are equal; a similar assertion holds for LDf. Finally, $f'(x)$ exists iff $LDf(x)$ and $RDf(x)$ exist and are equal.

Theorem 7.5.3   *Let f be a non-decreasing real-valued function on the interval* $[a, b]$. *Then*

1. *f is differentiable almost everywhere;*
2. *the derivative f' is measurable and;*
3. $\int_a^b f'(x)dx \le f(b) - f(a)$.

Proof:
(1) We will show first that the set where D_+, D^+, $D^- D_-$ disagree, has measure zero. Consider

$$E_-^+(p, q) = \{x \in (a, b) : D^+ f(x) > q > p > D_- f(x)\}$$

for each $p, q \in \mathbb{Q}$. Let $\alpha = \lambda(E_-^+(p, q))$ (verify that $E_-^+(p, q)$ is measurable first). Clearly $\alpha < \infty$. Let $\epsilon > 0$ be given. By Proposition 4.4.1, we can find an open set $U \subset \mathbb{R}^n$ such that $E_-^+(p, q) \subset U$ and $\lambda(U) < \alpha + \epsilon$.
   For each $x \in E_-^+(p, q)$, $\exists h(x)$ such that $I_h^x = (x - h, x + h) \subset U$ and

$$f(x) - f(x - h) < ph \tag{7.5.24}$$

for all $h \le h(x)$. Since $\mathcal{I} = \{I_h^x : x \in E_-^+(p, q), h < h(x)\}$ is a Vitali cover for $E_-^+(p, q)$, we may, by Lemma 7.1, find a sub-collection $\{I_i = (x_i - h_i, x_i + h_i) : i = 1, \ldots n\} \subset \mathcal{I}$ such that

$$\lambda(E_-^+(p, q) \cap \coprod_{i=1}^n I_i) > \alpha - \epsilon. \tag{7.5.25}$$

For each $y \in E_-^+(p, q) \cap \cup_{i=1}^n I_i$, there exists $k(y)$ such that $J_j^y = (y - j, y + j) \subset I_i$ for some $i = 1, \ldots, n$, and

$$f(y + j) - f(y) > qj, \tag{7.5.26}$$

for all $j \leq k(y)$. Since $\mathcal{J} = \{J_j^y : y \in E_-^+(p,q) \cap \cup_{i=1}^n I_i, j < k(y)\}$ is a Vitali cover for $E^+(p,q) \cap \cup_{i=1}^n I_i$, we may, by Lemma 7.1 and (7.5.25), find a sub-collection $\{J_1, J_2, \ldots, J_l\}$ such that

$$\lambda(E_-^+(p,q) \cap \coprod_{j=1}^l J_j) > \alpha - 2\epsilon. \tag{7.5.27}$$

Each interval $J_j, j = 1 \ldots l$ is contained in some interval I_i for some $i = 1, \ldots, n$. Thus,

$$p(\alpha + \epsilon) > p\lambda(U) \geq p\lambda \left(\coprod_{i=1}^n I_i \right) = p \sum_{k=1}^n h_k$$

$$> 2 \sum_{i=1}^n [f(x_i) - f(x_i - h_i)] \quad \text{(by (7.5.24))}$$

$$\geq 2 \sum_{k=1}^n \sum_{\{j:J_j \subset I_i\}} f(y_j+k_j) - f(y_j) \quad \text{(as } f \text{ is non-decreasing)}$$

$$= 2 \sum_{j=1}^l [f(y_j + k_j) - f(y_j)]$$

$$> q2 \sum_{i=1}^n k_i \quad \text{(by (7.5.26))}$$

$$= q\lambda \left(\coprod_{j=1}^l J_j \right) > q(\alpha - 2\epsilon) \quad \text{(by (7.5.27))}.$$

As $\epsilon > 0$ was arbitrary, this shows that $p\alpha \geq q\alpha$. Since $p < q$, we conclude that $\alpha = 0$. It now follows that the Lebesgue measure of $E_-^+ = \{x \in [a,b] : D^+ f(x) > D_- f(x)\}$ is zero. One can imitate the above argument to conclude the validity of the first sentence in the proof of this theorem (See Ex. 7.5.12).

This proves (1) as well as (2).

(3) For $x \geq b$, define $f(x) = f(b)$. Now, for $h \geq 0$, define $f_h : [a,b] \to \mathbb{R}_+$ as $f_h(x) = \frac{f(x+h)-f(x)}{h}$. Observe that f_h, f' are non-negative functions (by non-decreasing assumption on f) and hence Fatou's lemma applies to yield

$$\int_a^b f'(x)dx = \int_a^b \lim_{h \downarrow 0} f_h(x)dx$$

$$\leq \liminf_{h \downarrow 0} \int_a^b \frac{f(x+h) - f(x)}{h} dx$$

$$= \liminf_{h \downarrow 0} \frac{1}{h} \left[\int_b^{b+h} f(x)dx - \int_a^{a+h} f(x)dx \right]$$

$$= \liminf_{h \downarrow 0} \left[f(b) - \frac{1}{h} \int_a^{a+h} f(x)dx \right]$$

$$\leq f(b) - f(a).$$

The second last equality above follows because of the definition of f and the last inequality follows from the fact that f is increasing. This clearly establishes that f' is integrable, hence finite a.e. $\square$

Definition 7.5.4   *A collection of points $\pi = \{x_k : k = 0, 1, \ldots n\}$ is said to be a* **partition** *of $[a, b]$ if $a = x_0 \leq x_1 \leq \ldots \leq x_n = b$. The set of all partitions $[a, b]$ is denoted by $\mathcal{P}[a, b]$.*

Definition 7.5.5   *A function $f : [a, b] \to \mathbb{R}$ is said to be of* **bounded variation** *if*

$$T = \sup_{\pi \in \mathcal{P}([a,b])} \sum_{x_i \in \pi} | f(x_i) - f(x_{i-1}) |$$

is finite. In that case the quantity T is referred to as the **total variation** *of f.*

Theorem 7.5.6   *A function f is of bounded variation on $[a, b]$ if and only if f is the difference of two non-decreasing real-valued functions on $[a, b]$.*

Proof: Suppose $f = g - h$, where g, h are non-decreasing real-valued functions on $[a, b]$. Then, for any subdivision of $[a, b]$ we have,

$$\sum_{i=1}^n | f(x_i) - f(x_{i-1}) | \leq \sum_{i=1}^n | g(x_i) - g(x_{i-1}) | + \sum_{i=1}^n | h(x_i) - h(x_{i-1}) |$$

$$= g(b) - g(a) + h(b) - h(a).$$

Clearly, this implies that the total variation of f is finite.
    Conversely, for $x \in \mathbb{R}$ define

$$x^+ = x \vee 0, x^- = -(x \wedge 0).$$

Notice $|x| = x^+ + x^-, x = x^+ - x^-$. Hence if we let

$$P^a(x) := \sup_{\pi \in \mathcal{P}[a,x]} \sum_{i=1}^k (f(x_i) - f(x_{i-1}))^+,$$

$$N^a(x) := \sup_{\pi \in \mathcal{P}[a,x]} \sum_{i=1}^k -(f(x_i) - f(x_{i-1}))^-,$$

we find that P^a and N^a are non-negative bounded non-decreasing functions on $[a, b]$. Now let $\pi = \{a = x_1 \leq \ldots \leq x_k = b\} \in \mathcal{P}([a, x])$, then

$$f(x) - f(a) = \sum_{i=1}^{k} f(x_i) - f(x_{i-1})$$

$$= \sum_{i=1}^{k} (f(x_i) - f(x_{i-1}))^+ - \sum_{i=1}^{k} (f(x_i) - f(x_{i-1}))^-.$$

So,

$$\sum_{i=1}^{k} (f(x_i) - f(x_{i-1}))^+ = \sum_{i=1}^{k} (f(x_i) - f(x_{i-1}))^- + f(x) - f(a).$$

Taking suprema over all π, we have that $P^a(x) = N^a(x) + f(x) - f(a)$. So $f(x) = P^a(x) + f(a) - N^a(x)$ is the difference of two non-decreasing real-valued functions in $[a, b]$. □

Proposition 7.5.7 *Let* $f \in L^1([a, b], \lambda)$. *Then* $F(x) = \int_a^x f(y)dy$ *is differentiable a.e. and* $F'(x) = f(x)$ *a.e.*

Proof: Without loss of generality we shall assume that $f \geq 0$. As $f \in L^1([a, b], \lambda)$, we have that $F(x) = \int_a^x f(y)dy$ is a non-decreasing function and by Theorem 7.5.3, $F'(x)$ exists a.e. We consider two cases.

Case 1: $f \in L^\infty([a, b], \lambda)$
Let $c \in [a, b)$, then by dominated convergence

$$\int_a^c F'(x)dx = \int_a^c \left(\lim_{h \to 0} \frac{1}{h} \int_x^{x+h} f(y)dy \right) dx$$

$$= \lim_{h \to 0} \frac{1}{h} \int_a^c \int_x^{x+h} f(y)dydx$$

$$= \lim_{h \to 0} \frac{1}{h} \left(\int_c^{c+h} F(x)dx - \int_a^{a+h} F(x)dx \right).$$

We are interested in the limit h approaching zero, so we may assume h is small enough so that $a + h$ and $c + h$ belong to $[a, b]$. Now F is continuous (verify this). So if $u \in \{c, a\}$ and $\epsilon > 0$ then, for all sufficiently small h,

$$\left| \frac{1}{h} \int_u^{u+h} F(x)dx - F(u) \right| \leq \frac{1}{h} \int_u^{u+h} | F(x) - F(u) | \, dx$$

$$\leq \epsilon \frac{1}{h} \int_u^{u+h} dx = \epsilon.$$

As $F(a) = 0$, we have that

$$\int_a^c F'(x)dx = F(c) - F(a) = \int_a^c f(x)dx. \tag{7.5.28}$$

Since this is true for any $c \in [a, b)$, deduce from Ex. 2.3.6 that $F'(x) = f(x)$ a.e. $x \in [a, b]$.

Case 2: $f \in L^1([a, b], \lambda)$

Consider $f^K(x) = \min(f(x), K)$ for some $K > 0$. As $f - f^K \geq 0$, $G_K(x) = \int_a^x (f(y) - f^K(y))dy$ is a non-decreasing function and hence differentiable with non-negative derivative almost everywhere. So,

$$\frac{d}{dx}\left(\int_a^x f(y)dy \right) = \frac{d}{dx}\left(\int_a^x f(y) - f^K(y)dy \right) + \frac{d}{dx}\left(\int_a^x f^K(y)dy \right)$$

$$\geq \frac{d}{dx}\left(\int_a^x f^K(y)dy \right)$$

$$= f^K(x),$$

where the last equality is from Case 1 as f^K is bounded. Allowing $K \to \infty$, we find that

$$F'(x) \geq f(x) \text{ a. e.} \tag{7.5.29}$$

However, by Theorem 7.5.3 we have

$$\int_a^b F'(x)dx \leq F(b) - F(a) = \int_a^b f(x)dx. \tag{7.5.30}$$

Using (7.5.29), (7.5.30) and Theorem 2.3.4, we have that $F'(x) = f(x)$ a.e. $\square$

Definition 7.5.8 *A real-valued function f defined on $[a, b]$ is said to be* **absolutely continuous** *on $[a, b]$ if, given $\epsilon > 0$, there is $\delta > 0$ such that*

$$\sum_{i=1}^n | f(x_i) - f(y_i) | < \epsilon$$

for all finite collections of pairwise disjoint intervals (x_i, y_i) with $\sum_{i=1}^n (y_i - x_i) < \delta$.

Proposition 7.5.9 *Let $f : [a, b] \to \mathbb{R}$ be absolutely continuous.*

1. *Then f is of bounded variation and hence differentiable a.e.*
2. *If $f'(x) = 0$ a.e. in $[a, b]$, then f is a constant.*

Proof: Let $\epsilon > 0$ be given. Then there is $\delta > 0$ such that

$$\sum_{i=1}^{n} \mid f(x_i) - f(y_i) \mid < \epsilon \tag{7.5.31}$$

for all finite collection of disjoint intervals (x_i, y_i) with $\sum_{i=1}^{n}(y_i - x_i) < \delta$.

(1)   Choose N large enough so that we may divide the interval $[a, b]$ into N sub-intervals each of length less than δ. Then, clearly, the total variation T of f is less than or equal to $N\epsilon$. Therefore f is of bounded variation and consequently differentiable a.e. (by Theorems 7.5.6 and 7.5.3).

(2)   We shall show that $f(a) = f(c)$ for any $c \in [a, b]$. Let $E = \{x \in (a, c) : f'(x) = 0\}$. Clearly, $\lambda(E) = c - a$. Let $\eta > 0$ be given. For every $x \in E$, $\exists h(x)$ such that $I_h^x = (x - h, x + h) \subset (a, c)$ and

$$\mid f(x + h) - f(x - h) \mid < \eta h \tag{7.5.32}$$

for all $h \leq h(x)$. Since $\mathcal{I} = \{I_h^x : x \in E, h \leq h(x)\}$ is a Vitali cover of E, we may, by Lemma 7.1, find a finite collection of intervals $\{I_n, I_2, \ldots, I_n\} \in \mathcal{I}$ such that

$$\lambda \left(E - \coprod_{i=1}^{n} I_i \right) < \delta. \tag{7.5.33}$$

Let $I_k = (a_k, b_k)$ for $k = 1, 2 \ldots n$. By (7.5.31), (7.5.32), and (7.5.33), we may conclude the following inequalities:

$$\sum_{i=1}^{n} \mid f(b_i) - f(a_i) \mid \leq \eta \sum_{i=1}^{n}(b_i - a_i) < \eta(c - a), \tag{7.5.34}$$

$$\mid f(a) - f(a_1) \mid + \sum_{i=1}^{n-1} \mid f(a_{i+1}) - f(b_i) \mid + \mid f(c) - f(b_n) \mid < \epsilon. \tag{7.5.35}$$

Therefore,

$$| f(c) - f(a) |$$

$$=| f(a_1) - f(a) + \sum_{i=1}^{n}(f(b_i) - f(a_i)) + \sum_{i=1}^{n-1}(f(a_{i+1}) - f(b_i))$$

$$+ f(c) - f(b_n) |< \epsilon + \eta(c - a).$$

As $\epsilon, \eta > 0$ were arbitrary, we find that $f(c) = f(a)$.   □

Theorem 7.5.10   *Let $F : [a, b] \to \mathbb{R}$. The following conditions are equivalent:*

1.   *F is absolutely continuous.*
2.   *F is differentiable a.e., $F' \in L^1([a, b], \lambda)$ and*

$$F(x) = F(a) + \int_a^x F'(t)dt.$$

Proof: $(2) \Rightarrow (1)$ is immediate from Ex. 7.5.13.

$(1) \Rightarrow (2)$: Proposition 7.5.9 implies that F is differentiable a.e. and is of bounded variation. By Theorem 7.5.6. there exists two non-decreasing functions such that $F = F_1 - F_2$. Therefore, by Theorem 7.5.3 we have

$$\int_a^b | F'(x) | \, dx \leq \int_a^b | F_1'(x) - F_2'(x) | \, dx$$

$$\leq \int_a^b | F_1'(x) | \, dx + \int_a^b | F_2'(x) | \, dx$$

$$\leq F_1(b) - F_1(a) + F_2(b) - F_2(a).$$

Thus F' is integrable. So the function defined by $D(x) = F(x) - \int_a^x F'(y)dy$ is differentiable a.e. by Proposition 7.5.7. Further, $D'(x) = F'(x) - F'(x) = 0$ a.e. Hence by Proposition 7.5.9, D is a constant function. Therefore $D(x) = D(a)$. So (2) is proved.   □

EXERCISES

Ex. 7.5.11   Show that if f is integrable on $(\Omega, \mathcal{B}, \mu)$, then

$$\lim_{\mu(B) \to 0} \int_B f d\mu = 0,$$

(i.e, given any $\epsilon > 0$, there is $\delta > 0$ such that if $B \in \mathcal{B}$ and $\mu(B) < \delta$, then $| \int_B f d\mu | < \epsilon$).

Ex. 7.5.12   In the notation of Theorem 7.5.3, show that the sets E_+^-, E_-^-, E_+^+ have Lebesgue measure zero.

Ex. 7.5.13   Show that if $f \in L^1([a, b], \lambda)$, then $F(x) = \int_a^x f(y) dy$ is an absolutely continuous function.

Ex. 7.5.14   Let C be the Cantor set defined in Ex. 1.6.5.

1.   Show that if $x \in C$, then $x = \sum_{j=1}^{\infty} \frac{a_j}{3^j}$, where $a_j = 0$ or $a_j = 2$ for all j.
2.   Define a function $f : C \rightarrow [0, 1]$ as follows:

$$f(x) = \sum_{j=1}^{\infty} \frac{b_j}{2^j},$$

where $x = \sum_{j=1}^{\infty} \frac{a_j}{3^j}$ and $b_j = \frac{a_j}{2}$.

a)   Show that f maps C onto $[0, 1]$.
b)   If $x, y \in C$, $x < y$ and x, y are not the end points of one of the intervals removed from $[0, 1]$ to obtain C, then $f(x) < f(y)$.
c)   If $x, y \in C$, $x < y$ and x, y are end points of one of the intervals removed from $[0, 1]$ to obtain C, then show that $f(x) = f(y) = \frac{p}{2^k}$ for some $p, k \in \mathbb{N}$, p not divisible by 3. (*Hint*: If x is an end point of one of the intervals removed to obtain C, then $x = \frac{p}{3^k}$ for some $p, k \in \mathbb{N}$, p not divisible by 3. Use (1) and 2(a) to obtain the result.)
d)   Extend f to a map from $[0, 1]$ onto itself by defining its value on each interval missing from C to be its value at the end points. Show that f is continuous but not absolutely continuous (*Hint*: $f' = 0$ a.e.)

7.6   The Riesz Representation Theorem

This section is devoted to proving the Riesz representation theorem which asserts, loosely speaking, that 'positive linear functionals come from measures'. Thus, what we call the Riesz representation theorem is stated in three parts as Theorems 7.6.1, 7.6.7 and 7.6.9, corresponding to the compact metric, compact Hausdorff, and locally compact Hausdorff cases of the theorem. (Please refer to the appendix for the definitions and properties of such spaces.)

7.6.1   The compact metric case

In this section we shall prove a special and probably the most important case of the theorem, i.e., when the underlying space X is a compact metric space, which is our standing assumption throughout this section.

In this section, the symbol $\mathcal{B}_X$ will denote the *Borel σ-algebra of X*; i.e., $\mathcal{B}_X$ is the smallest σ-algebra of subsets of X which contains all compact sets.

Thus, our aim in this section is to prove the following fact.

Theorem 7.6.1 *If $\tau : C(X) \to \mathbb{C}$ is a linear functional which is positive (in the sense of assuming non-negative real values on non-negative real-valued continuous functions), then there exists a unique finite positive measure μ defined on the Borel σ-algebra $\mathcal{B}_X$ such that*

$$\tau(f) = \int f d\mu.$$

We prepare the way for the proof with a few simple results.

Lemma 7.6.2 *Let X be a compact Hausdorff space. Then the following conditions on a linear functional $\tau : C(X) \to \mathbb{C}$ are equivalent:*

(a) *τ is positive in the sense of the statement of the previous theorem;*
(b) *τ is a bounded linear functional and $\|\tau\| = \tau(1)$, where we write 1 for the constant function identically equal to one.*

Proof: This statement is known—see [1] for instance—to hold in the more general context where $C(X)$ is replaced by a general (not necessarily commutative) unital C^*-algebra. In the interest of a general readership, we present the proof in the special (commutative) case stated in this lemma.

$(a) \Rightarrow (b)$: Positivity of τ implies that the equation

$$(f, g) = \tau(f\bar{g})$$

defines a 'semi-inner product' on $C(X)$, i.e., the expression (f, g) is sesquilinear in its arguments (meaning $(\sum_{i=1}^2 \alpha_i f_i, \sum_{j=1}^2 \beta_j g_j) = \sum_{i,j=1}^2 \alpha_i \bar{\beta}_j (f_i, g_j)$) and positive semi-definite (meaning $(f, f) \geq 0$ for all f). It is a standard fact—found in any text in functional analysis, such as [25] or [26]—that every sesquilinear positive-semi-definite form satisfies the celebrated *Cauchy–Schwarz inequality*, i.e.,

$$|(f, g)|^2 \leq (f, f)(g, g) \ \forall \ f, g \in C(X) .$$

In other words,

$$|\tau(f\bar{g})|^2 \leq \tau(|f|^2)\tau(|g|^2) \ \forall \ f, g \in C(X) . \tag{7.6.36}$$

In particular, with $g = 1$, we find that

$$|\tau(f)|^2 \leq \tau(|f|^2)\tau(1).$$

But it follows from positivity that

$$\tau(|f|^2) \leq \tau(\|f\|^2 \, 1) = \|f\|^2 \tau(1).$$

Thus, we find that

$$|\tau(f)|^2 \leq \|f\|^2 \tau(1)^2 \; \forall \, f,$$

i.e., $\tau \in C(X)^*$ and $\|\tau\|_{C(X)^*} \leq \tau(1)$. The reverse inequality is obviously valid since $\|1\| = 1$; the proof of $(a) \Rightarrow (b)$ is complete.

$(b) \Rightarrow (a)$ We may assume, without loss of generality, that $\|\tau\| = \tau(1) = 1$. It will clearly suffice to show that

$$0 \leq f \leq 1 \Rightarrow 0 \leq \tau(f) \leq 1.$$

Suppose, conversely, that $\tau(f) = z \in \mathbb{C} \setminus [0, 1]$. Then we can find an open disc with centre z_0 and radius $r > 0$ say, which contains $[0,1]$ but not z. Then, for any $x \in X$, we have $f(x) \in [0, 1]$, so $|f(x) - z_0| < r$, and consequently $\|f - z_0 1\| < r$; hence $|z - z_0| = |\tau(f - z_0 1)| \leq \|f - z_0 1\| < r$. This contradiction to our assumption that $|z - z_0| \geq r$ completes the proof. □

Proposition 7.6.3 *Every compact metric space is a continuous image of the (compact metric) space $2^{\mathbb{N}}$, the Cartesian product of countably infinitely-many copies of a two-point space.*

Proof: The proof relies on two facts:

(i)   A compact metric space is totally bounded, i.e, for any $\epsilon > 0$ it is possible to cover X by finitely many balls with radius less than ϵ; and

(ii)   Cantor's intersection theorem which implies that the intersection in a compact metric space of a decreasing sequence of closed balls whose radii converge to zero, is a singleton set.

Indeed, by (i) above, we may find closed balls $\{B_k : 1 \leq k \leq n_1\}$ each with radius at most one, whose union is X. Next, since B_j is compact, we can find closed balls $\{B_{j,k} : 1 \leq k \leq m'_j\}$ of radius at most $\frac{1}{2}$ whose union contains B_j; choose $n_2 = \max_{1 \leq j \leq n_1} m'_j$, and define $B_{j,k} = B_{j,m'_j}$ for $m'_j \leq k \leq n_2$. In other words, we may assume that in the labelling of $B_{j,k}$, the range of the second index k is over the finite index set $\{1, \ldots, n_2\}$ and independent of the first index j. Using (i) repeatedly, an easy induction argument shows that we may, for each q, find a positive integer n_q and closed balls $\{B_{j_1, j_2, \ldots, j_q} : 1 \leq j_i \leq n_i\}$ of radius at most $\frac{1}{q}$ such that (a) $X = \cup_{j=1}^{n_1} B_j$, and (b) $B_{j_1, j_2, \ldots, j_{q-1}} \subset \cup_{j_q=1}^{n_q} B_{j_1, j_2, \ldots, j_q}$ for each $j_1, j_2, \ldots, j_{q-1}$.

If $\mathbf{j} = (j_1, j_2, \ldots, j_q, \ldots)$, appeal to (ii) above to find that $\cap_{q=1}^{\infty} B_{j_1, j_2, \ldots, j_q} = \{f(\mathbf{j})\}$ for a uniquely determined function $f : \prod_{q=1}^{\infty} \{1, 2, \ldots, n_q\} \to X$. The hypotheses ensure that if **i** and **j** agree in the first q coordinates, then $f(\mathbf{i})$ and $f(\mathbf{j})$ are at a distance of at most $\frac{2}{q}$ from one another. This shows that the function f is continuous. Finally, (a) and (b) of the last paragraph ensure that the function f maps onto X.

Finally, we may clearly assume without loss of generality that $n_q = 2^{m_q}$. Then, it is clear that there is a (continuous) bijection from $2^{\mathbb{N}}$ to $\prod_{q=1}^{\infty}\{1, 2, 3 \ldots, n_q\}$ (*Reason*: Take a bijection from $\mathbb{N}$ to $\coprod_{q=1}^{\infty}\{1, 2, \ldots, m_q\}$ and consider the induced mapping

$$2^{\mathbb{N}} \to 2^{\coprod_{q=1}^{\infty}\{1,2,\ldots,m_q\}} \equiv \prod_{q=1}^{\infty} 2^{\{1,2,3,\ldots,m_q\}} \equiv \prod_{q=1}^{\infty}\{1, 2, \ldots, n_q\}).$$

Combining these two maps we get the desired surjection to X.   □

Proof of Theorem 7.6.1

In addition to the above facts, the proof uses just one more fact, the Hahn–Banach theorem, which says that any bounded linear functional τ_0 on a subspace V_0 of a normed space V may be extended to a bounded linear functional τ on V such that $\|\tau\|_{V^*} = \|\tau_0\|_{V_0^*}$. This may be found in any text on functional analysis. (See [25] or [26], for instance.)

The first step of the proof is to observe that the special case of Theorem 7.6.1 when $X = 2^{\mathbb{N}}$, is a consequence of Caratheodory's extension theorem. Indeed, suppose τ is a positive linear functional on $C(2^{\mathbb{N}})$, which we may assume is normalised so that $\tau(1) = 1$. Let

$$\pi_{n]}(j_1, j_2 \ldots, j_n, \ldots) = (j_1, j_2 \ldots, j_n)$$

denote the projection $\pi_{n]} : 2^{\mathbb{N}} \to 2^n$ onto the first n coordinates (where we have written 2^n to denote the product of n copies of the two point space). Let $\mathcal{A} = \cup_{n=1}^{\infty}\mathcal{A}_{n]}$ where $\mathcal{A}_{n]} = \{\pi_{n]}^{-1}(E) : E \subset 2^n\}, n \in \mathbb{N}$. Then, $\mathcal{A}$ is a base for the topology of $2^{\mathbb{N}}$ and all the members of $\mathcal{A}$ are open and compact; $\mathcal{A}$ is an algebra of sets which generates the Borel σ-algebra $\mathcal{B}_{2^{\mathbb{N}}}$. In particular, the functions 1_A are *continuous* for each $A \in \mathcal{A}$. Hence, we may define

$$\mu(A) = \tau(1_A) \ \forall \ A \in \mathcal{A}. \tag{7.6.37}$$

The linearity of τ clearly implies that μ is a finitely additive set function on $\mathcal{A}$. Since members of $\mathcal{A}$ are compact and open, no element of $\mathcal{A}$ can be expressed as a countable union of pairwise disjoint non-empty members of $\mathcal{A}$. In other words, we see that any finitely additive set function on $\mathcal{A}$ is automatically (vacuously) countably additive, and hence, by Caratheodory's extension theorem, extends to a probability measure μ defined on all of $\mathcal{B}_X$. If we define $\tau_\mu(f) = \int f d\mu$, then we see from (7.6.37) that τ and τ_μ agree on any continuous function f which is $\mathcal{A}_{n]}$ measurable (and necessarily a simple function). Since such functions are dense in $C(2^N)$, we may conclude that $\tau = \tau_\mu$. In fact, (7.6.37) determines μ uniquely since a finite measure on $\mathcal{B}$ is uniquely determined by its restriction to any 'algebra of sets' which generates $\mathcal{B}$ as a σ-algebra. In other words, Theorem 7.6.1 is indeed true when $X = 2^N$.

Suppose now that X is a general compact metric space and τ is a positive linear functional on $C(X)$. We may assume that $\|\tau\| = \tau(1) = 1$. Proposition 7.6.3 guarantees the existence of a continuous surjection $p : 2^N \to X$. Then it is easy to see that the map $p^* : C(X) \to C(2^N)$ defined by $p^*(f) = f \circ p$ is an isometric positivity preserving homomorphism of algebras. In particular, we may regard $C(X)$ as a subspace of $C(2^N)$ via p^*. It follows from the Hahn–Banach theorem that there exists a bounded linear functional $\tilde{\tau}$ on $C(2^N)$ such that

$$\tilde{\tau}(p^*(f)) = \tau(f) \; \forall \; f \in C(X) \tag{7.6.38}$$

and

$$\|\tilde{\tau}\| = \|\tau\| = \tau(1) = \tilde{\tau}(p^*(1)) = \tilde{\tau}(1). \tag{7.6.39}$$

Deduce now from Lemma 7.6.2 that $\tilde{\tau}$ is a positive linear functional on $C(2^N)$. Conclude from the already proved special case of the theorem for 2^N that there exists a positive (in fact probability) measure ν defined on $\mathcal{B}_{2^N}$ such that $\tilde{\tau}(g) = \int_{2^N} g d\nu$. Hence, we see from the 'change of variable formula' (Proposition 7.4.1) that

$$\tau(f) = \tilde{\tau}(p^*(f))$$

$$= \int_{2^N} p^*(f) \, d\nu$$

$$= \int_{2^N} f \circ p \, d\nu$$

$$= \int_X f \, d(\nu \circ p^{-1})$$

$$= \int_X f \, d\mu,$$

with $\mu = \nu \circ p^{-1}$.

To complete the proof, we only need to establish uniqueness, For this, we first assert that if τ is given by integration against μ, then

$$\mu(K) = \inf\{\tau(f) : 1_K \le f \in C(X)\} \tag{7.6.40}$$

for every compact $K \subset X$.

(*Reason* : $1_K \le f \Rightarrow \mu(K) = \int 1_K d\mu \le \int f d\mu = \tau(f)$, it is clear that $\mu(K)$ is no greater than the infimum displayed in (7.6.40).

Conversely, since compact subsets of X are G_δ sets, we can find open sets U_n such that $K = \cap U_n$. (For instance, we can choose $U_n = \{x \in X : d(x, K) < \frac{1}{n}\}$.) By Theorem A.3.23 (see Appendix A.3), we can find continuous $f_n : X \to [0, 1]$ such that $1_K \le f_n \le 1_{U_n}$. Then it is clear that $f_n(x) \to 1_K(x) \; \forall x \in X$, and since $0 \le f_n \le 1 \; \forall n$, it follows from the dominated convergence theorem that

$$\mu(K) = \int 1_K d\mu = \lim_{n \to \infty} \int f_n d\mu,$$

and in particular, $\mu(K)$ is no smaller than the infimum displayed in (7.6.40).)

Finally, it follows from (7.6.40) and Proposition 1.7.2 that τ determines μ uniquely. $\qquad\square$

Remark 7.6.4   *Observe that if μ is a finite positive measure on the Borel σ-algebra $\mathcal{B}_X$ of a compact metric space X, then μ is automatically inner-regular, i.e.,*

$$\mu(E) = \sup\{\mu(K) : K \subset E, K \text{ compact}\}.$$

7.6.2   The general compact Hausdorff case

We assume in this section that X is any compact Hausdorff space. We begin with a couple of elementary lemmas.

Lemma 7.6.5

(a)   *The following conditions on a closed subset A of X are equivalent:*

  (i)   *A is a G_δ set, i.e., there exists a sequence $\{U_n\}$ of open sets such that $A = \cap_n U_n$;*
  (ii) *there exists a continuous function $f : X \to [0, 1]$ such that $A = f^{-1}(\{0\})$.*

Proof: (i) $\Rightarrow$ (ii): Compact Hausdorff spaces are normal; so we can, by Urysohn's lemma (Theorem A.3.23), find a continuous function $f_n : X \to [0, 1]$ such that

$$f_n(x) = \begin{cases} 0 & \text{if } x \in A \\ 1 & \text{if } x \notin U_n \end{cases}.$$

Now set $f = \sum_{n=1}^{\infty} 2^{-n} f_n$.

$(ii) \Rightarrow (i)$: Let $U_n = \{x \in X : |f(x)| < \frac{1}{n}\}$. $\square$

In this section, the symbol $\mathcal{B}_X$ will denote the the smallest σ-algebra of subsets of X which contains all the compact G_δ sets. A member of the σ-algebra $\mathcal{B}_X$ is called a *Baire set*. (Note that when X is a compact metric space, there is no distinction between Borel sets and Baire sets; this is because any closed set A in a metric space is a G_δ, as demonstrated by the function $f(x) = d(x, A)$.)

Lemma 7.6.6   *The following conditions on a subset $E \subset X$ are equivalent:*

(i)   *A is a Baire set.*

(ii)   *There exists a continuous function $F : X \to Y$ for X into a compact metric space Y, which may be assumed without loss of generality to be $[0, 1]^C$ for some countable set C, and a Baire (equivalently Borel) set $E \in \mathcal{B}_Y$ such that $A = F^{-1}(E)$.*

In particular, any continuous function from X to a compact metric space Y is 'Baire-measurable'; hence any scalar-valued continuous function on X is Baire-measurable.

Proof: $(ii) \Rightarrow (i)$: Suppose $F : X \to Y$ is a continuous function from X into a compact metric space Y. Let $\mathcal{C} = \{E \in \mathcal{B}_Y : F^{-1}(E)$ be a Baire set$\}$. First notice that if K is any closed set in Y, then K is also a G_δ set, and consequently $F^{-1}(K)$ is a compact G_δ set and hence a Baire set in X; hence $\mathcal{C}$ contains all closed sets in Y. Since the definition shows that $\mathcal{C}$ is clearly a σ-algebra, it follows that $\mathcal{C} = \mathcal{B}_Y$.

$(i) \Rightarrow (ii)$: Let $\mathcal{B}$ denote the collection of those subsets $A \subset X$ which are inverse images under continuous functions into Y (where $Y = [0, 1]^C$ for some countable set C) of Borel sets in Y. We check now that $\mathcal{B}$ is closed under countable unions. Suppose $A_n = F_n^{-1}(E_n)$ for some $E_n \in \mathcal{B}_{Y_n}$ and $n \in \mathbb{N}$, where $F_n : X \to Y_n = [0, 1]^{C_n}$ (with C_n some countable set) is continuous; we may assume without loss of generality that the index sets C_n are pairwise disjoint. Then, let $C = \coprod_{n=1}^\infty C_n$ so that we may identify $Y = [0, 1]^C$ with $\prod_{n=1}^\infty Y_n$. Write π_n for the natural projection mapping of Y onto Y_n. Define $F : X \to Y$ by requiring that $\pi_n \circ F = F_n$, and define $G_n = \pi_n^{-1}(E_n)$. The definitions show that F is continuous, $G_n \in \mathcal{B}_Y$, so that $\cup_n G_n \in \mathcal{B}_Y$ and

$$\cup_n A_n = \cup_n F^{-1}(G_n) = F^{-1}(\cup_n G_n);$$

so $\mathcal{B}$ is indeed closed under countable unions. Since it is trivially closed under complementation, it follows that $\mathcal{B}$ is a σ-algebra of sets. So, in order to complete the proof of the lemma, it is to be proved that $\mathcal{B}$ contains all compact G_δ sets; but this is guaranteed by Lemma 7.6.5. $\square$

We are now ready for the Riesz representation theorem for general compact Hausdorff spaces.

Theorem 7.6.7 *If $\tau : C(X) \to \mathbb{C}$ is a positive linear functional, then there exists a unique finite positive measure μ defined on the Baire σ-algebra $\mathcal{B}_X$ such that*

$$\tau(f) = \int f d\mu.$$

Proof: Let $\mathcal{C} = \{(Y, \pi) : Y \text{ is metrisable and } \pi \text{ is a continuous surjection of } X \text{ onto } Y\}$.

For $(Y, \pi), (Y_1, \pi_1), (Y_2, \pi_2) \in \mathcal{C}$, let us write

$$\mathcal{B}_\pi = \{\pi^{-1}(E) : E \in \mathcal{B}_Y\}, \text{ and}$$

$$(Y_1, \pi_1) \le (Y_2, \pi_2) \Leftrightarrow \exists \, \psi : Y_2 \to Y_1 \text{ such that } \pi_1 = \psi \circ \pi_2.$$

The proof involves a series of assertions:

(1)   If $(Y, \pi) \in \mathcal{C}$, then there exists a unique measure μ_Y defined on $\mathcal{B}_Y$ such that

$$\int_Y g \, d\mu_Y = \tau(g \circ \pi), \ \forall \, g \in C(Y) . \tag{7.6.41}$$

(*Reason:* The equation

$$\tau_Y(g) \ = \ \tau(g \circ \pi) \tag{7.6.42}$$

defines a positive linear functional on $C(Y)$ with Y a compact metric , and we may appeal to Theorem 7.6.1.)

(2)   If $(Y_1, \pi_1), (Y_2, \pi_2) \in \mathcal{C}$ and if $(Y_1, \pi_1) \le (Y_2, \pi_2)$ so that there exists a continuous map $\psi : Y_2 \to Y_1$ such that $\pi_1 = \psi \circ \pi_2$, then $\mu_{Y_1} = \mu_{Y_2} \circ \psi^{-1}$.

(*Reason:* For arbitrary $g \in C(Y_1)$, an application of the 'change of variable formula' shows that

$$\int_{Y_1} g \, d(\mu_{Y_2} \circ \psi^{-1}) = \int_{Y_2} g \circ \psi \, d\mu_{Y_2}$$

$$= \tau(g \circ \psi \circ \pi_2)$$

$$= \tau(g \circ \pi_1) ,$$

and the uniqueness assertion of (1) above establishes the desired equality of measures.)

(3)   If $\{(Y_n, \pi_n) : n \in \mathbb{N}\} \subset C$, then there exists $(Y, \pi) \in C$ such that $(Y_n, \pi_n) \leq (Y, \pi)$ for $n \in \mathbb{N}$.

(*Reason:* Define $\pi : X \to \prod_{n=1}^{\infty} Y_n$ by $\pi(x) = (\pi_n(x))_{n=1}^{\infty}$, set $Y = \pi(X)$, and let $\psi_n : Y \to Y_n$ be the restriction to Y of the projection onto Y_n for $n \in \mathbb{N}$.)

Finally, it is a direct consequence of Lemma 7.6.6 that $\mathcal{B}_X = \cup_{(Y,\pi) \in C} \mathcal{B}_\pi$, while the above assertions (1)–(3) show that there exists a unique, well-defined finite measure μ on $\mathcal{B}_X$ with the property that $\mu(\pi^{-1}(E)) = \mu_Y(E)$ whenever $E \in \mathcal{B}_Y$; in other words, $\mu \circ \pi^{-1} = \mu_Y$ so $\int_X g \circ \pi \, d\mu = \int_Y g \, d\mu_Y = \tau(g \circ \pi) \forall g \in C(Y)$. Now if $f \in C(X)$, set $Y = f(X)$, $\pi = f$, define $g \in C(Y)$ by $g(z) = z \, \forall z \in Y$ and deduce from the previous sentence that indeed

$$\int_X f \, d\mu = \tau(f).$$

as desired. The proof of uniqueness of μ is exactly as in the compact metric case, except that the appeal in that proof to Proposition 1.7.2 must be replaced by an appeal to Remark 7.6.8 below. So the proof of Theorem 7.6.7 is complete.   □

Remark 7.6.8   *Observe that if μ is a finite positive measure on the Baire σ-algebra $\mathcal{B}_X$ of a compact Hausdorff space X, then μ is automatically inner-regular, i.e.,*

$$\mu(A) = \sup\{\mu(K) : K \subset A, \, K \text{ compact } G_\delta\}.$$

(*Reason:* Indeed, if A is any Baire subset of X, then there exists a compact metric space Y, a Borel subset $E \subset Y$ and a continuous map $F : X \to Y$ such that $A = F^{-1}(E)$. Define $\nu = \mu \circ F^{-1}$ on $\mathcal{B}_Y$. By Remark 7.6.4, we can find compact sets $C_n \subset E$ such that $\nu(E) = \sup_n \nu(C_n)$. Notice that $K_n = F^{-1}(C_n)$ is a compact G_δ subset of X such that $K_n \subset A$ and conclude that

$$\mu(A) = \nu(E) = \sup_n \nu(C_n) = \sup_n \mu(K_n).)$$

7.6.3   The locally compact case

In this section, we assume that X_0 is a locally compact Hausdorff space. We shall write

$$C_c(X_0) = \{f : X_0 \to \mathbb{C} : f \text{ continuous, supp}(f) \text{ compact}\} ,$$

where we write 'supp(f)' to denote the support of f, i.e., the closure of $\{x \in X_0 : f(x) \neq 0\}$. As before, we shall let $\mathcal{B}_{X_0}$ denote the σ-algebra generated by compact G_δ subsets of X_0.

Let us call a positive measure μ defined on $\mathcal{B}_{X_0}$ *inner regular* if:

(i) $\mu(K) < \infty$ for every compact G_δ subset of X_0, and
(ii) $\mu(E) = \sup\{\mu(K) : K$ a compact G_δ subset of $E\}$ for all $E \in \mathcal{B}_X$.

We now wish to derive the following result from the preceding sections.

Theorem 7.6.9   *If $\tau : C_c(X_0) \to \mathbb{C}$ is a positive linear functional, then there exists a unique positive inner regular measure μ defined on $\mathcal{B}_{X_0}$ such that*

$$\tau(f) = \int f d\mu \ \forall \ f \in C_c(X_0) \,.$$

Proof: Let K be a compact G_δ subset of X_0, and suppose $K = \cap_n U_n$, with U_n open. We may, and do assume without loss of generality that U_n has compact closure and that $\overline{U_{n+1}} \subset U_n$. For each n, we may, by Ex. 7.6.10, pick a continuous function $\phi_n : X \to [0, 1]$ such that $1_{\overline{U_{n+1}}} \leq \phi_n \leq 1_{U_n}$, i.e.,

$$\phi_n(x) = \begin{cases} 0 & \text{if } x \notin U_n \\ 1 & \text{if } x \in \overline{U_{n+1}} \end{cases} .$$

The construction implies that

$$0 \leq \phi_{n+1} = \phi_n \phi_{n+1} \leq \phi_n. \tag{7.6.43}$$

Assertion: Suppose now that $f \in C(K)$ and $f \geq 0$. Suppose[2] $\widetilde{f} \in C_c(X_0)$ is a non-negative extension of f, i.e., $\widetilde{f}|_K = f$. Then $\lim_{n \to \infty} \tau(\widetilde{f} \phi_n)$ exists and this limit depends only on f and is independent of the choices of any of $U_n, \phi_n, \widetilde{f}$.

Reason: To start with, the positivity of τ and (7.6.43) imply that $\{\tau(\widetilde{f} \phi_n) : n = 1, 2, \dots\}$ is a non-increasing sequence of non-negative numbers, and hence converges. If g is another continuous non-negative function with compact support which extends f, and if $\epsilon > 0$, then $\overline{U_n} \cap \{x \in X_0 : |\widetilde{f}(x) - g(x)| \geq \epsilon\}$ is seen to be a a decreasing sequence of compact sets whose intersection is empty. So, there exists an n such that

$$x \in U_n \Rightarrow |\widetilde{f}(x) - g(x)| < \epsilon;$$

it follows from the positivity of τ that

$$|\tau(\widetilde{f} \phi_k) - \tau(g \phi_k)| < \epsilon \tau(\phi_k)$$

for all large k. Since $\lim_{n \to \infty} \tau(\phi_n)$ exists (by virtue of the conclusion of the previous paragraph applied to $f = 1_K$, with $\widetilde{f} = 1_{\overline{U_1}}$ (say)), this establishes that the limit of the assertion is indeed independent of $\widetilde{f}$.

---

[2]Such an extension exists by Ex. 7.6.11

Suppose $\{(V_n, \psi_n)\}_n$ is an alternative choice to $\{(U_n, \phi_n)\}_n$ in the sense that (i) V_n is a sequence of open sets such that $K = \cap_n V_n$, (ii) $\psi_n \in C_c(X_0)$, $\psi_n : X_0 \to [0, 1]$ and (iii)

$$\psi_n(x) = \begin{cases} 0 & \text{if } x \notin V_n \\ 1 & \text{if } x \in V_{n+1} \end{cases}.$$

Then, it may be seen, exactly as above, that if $\epsilon > 0$ is arbitrary, then $\sup_{x \in X_0} |\tilde{f}(x)(\phi_n(x) - \psi_n(x))| < \epsilon$ for large n, and hence it is deduced from the positivity of τ that

$$\lim_{n \to \infty} \tau(\tilde{f}\phi_n) = \lim_{n \to \infty} \tau(\tilde{f}\psi_n),$$

thereby completing the proof of the assertion.

It is clear that there exists a linear functional τ_K on $C(K)$ such that

$$\tau_K(f) = \lim_n \tau(\tilde{f}\phi_n)$$

for any non-negative f (and $\tilde{f}$ as above). Since τ_K is a positive linear functional on $C(K)$, we may deduce from Theorem 7.6.7 that there exists a unique **finite positive** measure μ_K defined on $\mathcal{B}_K$ such that

$$\tau_K(f) = \int_K f \, d\mu_K \, \forall \, f \in C(K).$$

Notice that the collection Λ of compact G_δ sets is a directed set with respect to the ordering defined by

$$K \leq L \iff K \subset \text{Int}(L),$$

where $\text{Int}(L)$ denotes the interior of L.

We wish now to show that the family $\{\mu_K : K \in \Lambda\}$ is consistent in the sense that

$$K, L \in \Lambda, K \leq L \Rightarrow \mu_L|_K = \mu_K. \tag{7.6.44}$$

First notice that it follows from the definitions that if $g \in C_c(X_0)$ and if $\text{supp}(g) \leq L$, then

$$\int_L g \, d\mu_L = \tau(g). \tag{7.6.45}$$

Suppose now that $K, L \in \Lambda$ and $K \leq L$. We may find a sequence $\{U_n, \phi_n\}$ as above such that $K = \cap_n U_n$ and $U_1 \subset L$. Now, if $f \in C(K)$ is arbitrary we may, by Ex. 7.6.10 and 7.6.11, find an extension $\tilde{f}$ of f with $\mathrm{supp}(\tilde{f}) \leq L$. Deduce from (7.6.45) and the dominated convergence, for instance, that

$$\int_K f \, d\mu_K = \lim_n \tau(\tilde{f}\phi_n)$$

$$= \lim_n \int_L (\tilde{f}\phi_n) \, d\mu_L$$

$$= \int 1_K f \, d\mu_L$$

$$= \int_K f \, d(\mu_L)|_K .$$

The uniqueness assertion in Theorem 7.6.7 then shows that (7.6.44) is indeed valid.

Finally, for any set $E \in \mathcal{B}_{X_0}$, notice that $\{\mu_K(E \cap K) : K \in \Lambda\}$ is a non-decreasing set of real numbers which must converge to a number—call it $\mu(E)$ - in $[0, \infty]$. Note that

$$\mu(E) = \sup_{K \in \Lambda} \mu_K(E \cap K) \, \forall E \in \mathcal{B}_{X_0}. \tag{7.6.46}$$

We wish to say that this μ is the measure on $\mathcal{B}_{X_0}$ whose existence is asserted by the theorem. Suppose $E = \coprod_{n=1}^{\infty} E_n$, where $E, E_n \in \mathcal{B}_{X_0}$. Then

$$\mu(E) = \sup_{K \in \Lambda} \mu_K(E \cap K)$$

$$= \sup_{K \in \Lambda} \mu_K \left(\coprod_{n=1}^{\infty} E_n \cap K \right)$$

$$= \sup_{K \in \Lambda} \sum_{n=1}^{\infty} \mu_K(E_n \cap K)$$

$$= \sup_{K \in \Lambda} \sup_{N} \sum_{n=1}^{N} \mu_K(E_n \cap K)$$

$$= \sup_{N} \sup_{K \in \Lambda} \sum_{n=1}^{N} \mu_K(E_n \cap K)$$

$$= \sup_{N} \lim_{K \in \Lambda} \sum_{n=1}^{N} \mu_K(E_n \cap K)$$

$$= \sup_{N} \sum_{n=1}^{N} \lim_{K \in \Lambda} \mu_K(E_n \cap K)$$

$$= \sup_N \sum_{n=1}^{N} \mu(E_n)$$

$$= \sum_{n=1}^{\infty} \mu(E_n),$$

and hence μ is indeed countably additive, i.e., a measure defined on $\mathcal{B}_{X_0}$.

Now, if $f \in C_c(X_0)$, it follows from (7.6.45) and (7.6.44) that if $L \in \Lambda$ satisfies $\operatorname{supp}(f) \leq L$, then

$$\tau(f) = \int_L f \, d\mu_L = \int_{X_0} f \, d\mu.$$

Finally, if K is any compact G_δ, then it is clear that

$$\mu(K) = \mu_K(K) < \infty.$$

If $E \in \mathcal{B}_{X_0}$ is arbitrary, we have, by definition of our μ and Remark 7.6.8,

$$\mu(E) = \sup_{K \in \Lambda} \mu_K(E \cap K)$$

$$= \sup_{K \in \Lambda} \sup\{\mu_K(C) : C \subset E \cap K, \ C \text{ compact } G_\delta\}$$

$$= \sup\{\mu(C) : C \subset E, \ C \text{ compact } G_\delta\}. \tag{7.6.47}$$

Hence μ is inner-regular.

If μ_1 also represents τ as in the theorem, then we claim that

$$\mu_1(K) = \inf\{\tau(f) : 1_K \leq f \in C_c(X_0)\}$$

for all compact G_δ subsets K.(This is proved exactly as in the proof of (7.6.40).) Hence the requirement of inner-regularity of the representing measure forces $\mu_1 = \mu$ and the proof of the theorem is finally complete.   □

EXERCISES

Ex. 7.6.10   Let X be a locally compact Hausdorff space. If $K \subset X$ is compact and $A \subset X$ is closed such that $A \cap X = \emptyset$, then there is a $f \in C_c(X)$ such that $1_K \leq f \leq 1_{A'}$. (*Hint:* Consider the one-point compactification $\hat{X} = X \cup \{\infty\}$. Appeal to local compactness of X to find an open set U with compact closure in X such that $K \subset U \subset X \backslash A$. Find $f \in C(\hat{X})$ such that $1_K \leq f \leq 1_{\hat{X} \backslash U}$.)

Ex. 7.6.11   Let X be a locally compact Hausdorff space. If $K \subset X$ is compact and $f \in C(K)$, then there exists $g \in C_c(X)$ such that $g|_K = f$. (*Hint*: Let $L = \bar{U}$ in the notation for the hint of the previous exercise. Appeal to Tietze's extension theorem (Theorem A.3.24–see Appendix A.3) to find $f_1 \in C(L)$ such that $f_1|K = f$. By the previous exercise, find a $\phi \in C_c(X)$ such that $1_K \leq \phi \leq 1_{U'}$; set $g(x) = f_1(x)\phi(x)$ for $x \in L$ and $g(x) = 0$ otherwise.)

A

Appendix

A.1 Metric Spaces

Given a set X, a function $d : X \times X \to [0, \infty)$ which satisfies the following conditions for all $x, y, z \in X$ is called a **metric** on X:

$$d(x, y) = 0 \Leftrightarrow x = y \qquad\qquad (A.1.1)$$

$$d(x, y) = d(y, x) \qquad\qquad (A.1.2)$$

$$d(x, y) \leq d(x, z) + d(z, y), \qquad\qquad (A.1.3)$$

and the pair (X, d) is called a **metric space**. We shall sometimes be sloppy and simply refer to X as a metric space.

The quantity $d(x, y)$ is to be thought of as the distance between the points x and y. Thus, the third condition on the metric is the familiar *triangle inequality*.

The most familiar example, of course, is $\mathbb{R}^3$, with the metric defined by

$$d((x_1, x_2, x_3), (y_1, y_2, y_3)) = \sqrt{\sum_{i=1}^{3} (x_i - y_i)^2}.$$

Any one metric space gives rise to many metric spaces thus: if Y is a subset of X, then $(Y, d|_{Y \times Y})$ is also a metric space and is referred to as a (metric) subspace of X. Thus, for instance, the unit sphere $S^2 = \{x \in \mathbb{R}^3 : d(x, 0) = 1\}$ is a metric space; clearly there are many more interesting metric subspaces of $\mathbb{R}^3$. It is customary to write $\|x\| = d(x, 0)$, refer to this quantity as the **norm** of x and to think of S^2 as the set of vectors in $\mathbb{R}^3$ of unit norm.

The set $\mathbb{R}^3$ is an example of a set which is simultaneously a metric space as well as a vector space (over the field of real numbers), where the metric *arises from a norm*. We will be interested in more such general objects, which we now pause to define.

Definition A.1.1 *A norm on a vector space V is a function $V \ni x \mapsto \|x\| \in [0, \infty)$ which satisfies the following conditions for all $x, y \in V, \alpha \in \mathbb{C}$:*

$$\|x\| = 0 \Leftrightarrow x = 0 \tag{A.1.4}$$

$$\|\alpha x\| = |\alpha| \|x\| \tag{A.1.5}$$

$$\|x + y\| \le \|x\| + \|y\| \tag{A.1.6}$$

and a **normed vector space** *is a pair* $(V, \|\cdot\|)$ *consisting of a vector space and a norm on it.*

Example A.1.2

(1)   It is a fact that ℓ_n^p for any $n \in \mathbb{N}$ is a normed vector space, where $\ell_n^p = \mathbb{C}^n$ is a set and the norm is defined by

$$\|\alpha\|_p = \left(\sum_k |\alpha_k|^p \right)^{\frac{1}{p}}. \tag{A.1.7}$$

This is a special case of the more general $L^p(\Omega, \mathcal{B}, \mu)$ spaces considered in Section 7.2.

(2)   Given a non-empty set X, let $B(X)$ denote the space of bounded complex-valued functions on X. It is not hard to see that $B(X)$ becomes a normed vector space with respect to the norm (denoted by $\|\cdot\|_\infty$) defined by

$$\|f\|_\infty = \sup\{|f(x)| : x \in X\}. \tag{A.1.8}$$

Readers who are unfamiliar with some of the notions discussed below might care to refer to Section A.3 and Section A.4 of the Appendix where they will find definitions of compact and locally compact spaces respectively, while also having have the pleasure of making the acquaintance of 'functions vanishing at infinity'.

It should be noted that if X is a compact Hausdorff space, then the set $C(X)$ of continuous complex-valued functions on X is a vector subspace of $B(X)$ (since continuous functions on X are necessarily bounded), and is consequently a normed vector space in its own right.

More generally, if X is a locally compact Hausdorff space, let $C_0(X)$ denote the space of all complex-valued continuous functions on X which 'vanish at infinity'; thus $f \in C_0(X)$ precisely when $f : X \to \mathbb{C}$ is a continuous function such that for any positive ϵ, it is possible to find a compact subset $K \subset X$ such that $|f(x)| < \epsilon$ whenever $x \notin K$. The reader should verify that $C_0(X) \subset B(X)$—in fact it is the closure in $B(X)$ of the space of compactly supported comtinuous functions—and hence that $C_0(X)$ is a normed vector space with respect to $\|\cdot\|_\infty$. □

A.2   Topological Spaces

A topological space is a set X where there is a notion of 'nearness' of points (although there is no metric in the background) and (experience shows) the natural context to discuss notions of continuity. Thus, to each point $x \in X$, there is singled out some family $\mathcal{N}(x)$ of subsets of X whose members are called 'neighbourhoods of the point x'. (If X is a metric space, we say that N is a neighbourhood of a point x if it contains all points sufficiently close to x, i.e., if there exists some $\epsilon > 0$ such that $N \supset B(x, \epsilon) = \{y \in X : d(x, y) < \epsilon\}$.) A set is thought of as being 'open' if it is a neighbourhood of each of its points, i.e., U is open if and only if $U \in \mathcal{N}(x)$ whenever $x \in U$. A topological space is an axiomatisation of this setup; what we find is a simple set of requirements (or axioms) to be met by the family, call it τ, of all those sets which are open in the sense above.

Definition A.2.1   *A **topological space** is a set X equipped with a distinguished collection τ of subsets of X, the members of τ being referred to as **open** sets, such that the following axioms are satisfied:*

(1)   *$X, \emptyset \in \tau$; thus the universe of discourse and the empty set are assumed to be open;*

(2)   *$U \ V \in \tau \Rightarrow U \cap V \in \tau$ (and consequently, any finite intersection of open sets is open); and*

(3)   *$\{U_i : i \in I\} \subset \tau \Rightarrow \cup_{i \in I} U_i \in \tau$.*

*The family τ is called the **topology** on X. A subset $F \subset X$ will be said to be closed precisely when its complement $X - F$ is open.*

Thus a topology on X is just a family of sets which contains the whole space and the empty set, and which is closed under the formation of finite intersections and arbitrary unions. Of course, every metric space is a topological space in a natural way. Thus, we declare that a set is open precisely when it is expressible as a union of open balls. Some other easy (but perhaps somewhat pathological) examples of topological spaces are given below.

Example A.2.2

(1)   Let X be any set; define $\tau = \{X, \emptyset\}$. This is a topology called the **indiscrete topology**.

(2)   Let X be any set; define $\tau = 2^X = \{A : A \subset X\}$; thus, every set is open in this topology and this is called a **discrete topology**.

(3)   Let X be any set; the so-called '**co-finite topology**' is defined by declaring a set to be open if it is either empty or if its complement is finite.

(4)   Replacing every occurrence of the word 'finite' in (3) above by the word 'countable' gives rise to a topology called (naturally) the **'co-countable topology'** on X. (Of course, this topology would be interesting only if the set X is uncountable, just as the co-finite topology would be interesting only for infinite sets X since in the contrary case the resulting topology would degenerate into discrete topology.)   □

It is sometimes necessary to use the notion of *nets*, which we pause to define and elucidate with some examples.

Definition A.2.3   *A **directed set** is a partially ordered set $(I, \leq)$ which, in addition to the usual requirements of reflexivity $(i \leq i)$, antisymmetry $(i \leq j, j \leq i \Rightarrow i = j)$ and transitivity $(i \leq j, j \leq k \Rightarrow i \leq k)$ that have to be satisfied by any partially ordered set, satisfies the following property:*

$$\forall\, i, j \in I,\ \exists\, k \in I \ such\ that\ i \leq k, j \leq k.$$

*A **net** in a set X is a family $\{x_i : i \in I\}$ of elements of X which is indexed by some directed set I.*

Example A.2.4

(1)   The motivating example for the definition of a directed set is the set $\mathbb{N}$ of natural numbers with the natural ordering, and a net indexed by this directed set is nothing but a sequence.

(2)   Let S be any set and let $\mathcal{F}(S)$ denote the class of all finite subsets of S. This is a directed set with respect to the order defined by inclusion; thus, $F_1 \leq F_2 \Leftrightarrow F_1 \subseteq F_2$. This will be referred to as the directed set of finite subsets of S.

(3)   Let X be any topological space, and let $x \in X$; let $\mathcal{N}(x)$ denote the family of all open neighbourhoods of the point x; then $\mathcal{N}(x)$ is directed by 'reverse inclusion', i.e., it is a directed set if we define $U \leq V \Leftrightarrow V \subseteq U$.

(4)   If I and J are directed sets, then the Cartesian product $I \times J$ is a directed set with respect to the 'coordinate-wise ordering' defined by $(i_1, j_1) \leq (i_2, j_2) \Leftrightarrow i_1 \leq i_2$ and $j_1 \leq j_2$.   □

The reason that nets were introduced in the first place was to have analogues of sequences when dealing with more general topological spaces than metric spaces. In fact, if X is a general topological space, we say that a net $\{x_i : i \in I\}$ in X converges to a point x if, for every open neighbourhood U of x, it is possible to find an index $i_0 \in I$ with the property that $x_i \in U$ whenever $i_0 \leq i$. (As with sequences, we shall, in the sequel, write $i \geq j$ to mean $j \leq i$ in an abstract partially ordered set.)

Ex. A.2.5

(1) If $f : X \to Y$ is a map of topological spaces and if $x \in X$, then show that f is continuous at x if and only the net $\{f(x_i) : i \in I\}$ converges to $f(x)$ in Y whenever $\{x_i : i \in I\}$ is a net which converges to x in X. (*Hint*: Use the directed set of Example A.2.4(3).)

(2) Recall that a sequence $\{x_n : n \in \mathbb{N}\}$ in a metric space is said to be a Cauchy sequence if for each $\epsilon > 0$ there is an $N \in \mathbb{N}$ such that $d(x_n, x_m) < \epsilon$ for all $n, m \geq N$. Define what should be meant by saying that a net in a metric space is a 'Cauchy net', and show that every convergent net is a Cauchy net.

(3) Recall that a metric space is said to be complete if every Cauchy sequence converges, and then show that every Cauchy net in X also converges if X is complete. (*Hint*: For each n, pick $i_n \in I$ such that $i_1 \leq i_2 \leq \cdots \leq i_n \leq \cdots$, and such that $d(x_i, x_j) < \frac{1}{n}$, whenever $i, j \geq i_n$; now show that the net should converge to the limit of the Cauchy sequence $\{x_{i_n}\}_{n \in \mathbb{N}}$.)

(4) Is the Cartesian product of two directed sets directed with respect to the 'dictionary ordering'?

Just as convergent sequences are a useful notion when dealing with metric spaces, nets (see Definition A.2.3 as well as Example A.2.4(3)) will be very useful while dealing with general topological spaces.

Proposition A.2.6 *The following conditions on a subset F of a topological space are equivalent, and such sets are said to be* **closed***:*

(i) $U = X - F$ *is open;*

(ii) *if* $\{x_i : i \in I\}$ *is any net in F which converges to a limit* $x \in X$, *then* $x \in F$.

Proof: $(i) \Rightarrow (ii)$: Suppose $x_i \to x$ as in (ii) and suppose $x \notin F$, then $x \in U$, and since U is open, the definition of a convergent net shows that there exists some $i_0 \in I$ such that $x_i \in U \; \forall i \geq i_0$; but this contradicts the assumption that $x_i \in F \; \forall i$.

$(ii) \Rightarrow (i)$: Suppose (ii) is satisfied; we assert that if $x \in U$, then there exists an open set $B_x \subset U$ such that $x \in B_x$. (This will exhibit U as the union $\bigcup_{x \in U} B_x$ of open sets and thereby establish (i).) If our assertion was false, there must be a point $x \in U$ such that for every open neighbourhood V of x, i.e., an open set containing x, it is possible to find a point $x_V \in V \cap F$. Then, (see Example A.2.4(3)) $\{x_V : V \in \mathcal{N}(x)\}$ is a net in F which converges to the point $x \notin F$; the desired contradiction has been reached.   □

We gather a few more simple facts concerning closed sets in the next result.

Proposition A.2.7   *Let X be a topological space. Let us temporarily write $\mathcal{F}$ for the class of all closed sets in X.*

(1)   *The family $\mathcal{F}$ has the following properties (and a topological space can clearly be equivalently defined as a set where there is a distinguished class $\mathcal{F}$ of subsets of X which are called closed sets and which satisfy the following properties):*

(a) $X, \emptyset \in \mathcal{F};$

(b) $F_1, F_2 \in \mathcal{F} \Rightarrow (F_1 \cup F_2) \in \mathcal{F};$

(c)   *if I is an arbitrary set, then* $\{F_i : i \in I\} \subset \mathcal{F} \Rightarrow \cap_{i \in I} F_i \in \mathcal{F}.$

(2)   *if $A \subset X$ is any subset of X and the* **closure** *of A, denoted by the symbol $\overline{A}$, is the set defined by*

$$\overline{A} = \cap \{F \in \mathcal{F} : A \subset F\}, \tag{A.2.9}$$

then

(a) $\overline{A}$ *is a closed set and the smallest closed set which contains A;*

(b) $A \subset B \Rightarrow \overline{A} \subset \overline{B};$

(c) $x \in \overline{A} \Leftrightarrow U \cap A \neq \emptyset$ *for every open set containing x.*

Proof: The proof of (1) is a simple exercise in complementation and uses nothing more than 'de Morgan's laws'.

(2) (a) is a consequence of (1)(c), while (b) follows immediately from (a); (c) follows from the definitions and the fact that a set is closed precisely when its complement is open.   □

The following exercise lists some simple consequences of the definition of the closure operation and also contains the definition of a very important notion.

Ex. A.2.8

(1)   Let A be a subset of a topological space X and let $x \in A$. Then show that $x \in \overline{A}$ if and only if there is a net $\{x_i : i \in I\}$ in A such that $x_i \to x$.

(2)   If X is a metric space, show that nets can be replaced by sequences in (1).

(3)   A subset D of a topological space is said to be **dense** if $\overline{D} = X$. (More generally, a set A is said to be 'dense in a set B' if $B \subset \overline{A}$.) Show that the following conditions on the subset $D \subset X$ are equivalent:

(i) D is dense (in X);

(ii)   for each $x \in X$, there exists a net $\{x_i : i \in I\}$ in D such that $x_i \to x$;

(iii) if U is any non-empty open set in X, then $D \cap U \neq \emptyset$.

In dealing with metric spaces, we rarely have to deal explicitly with general open sets; open balls suffice in most contexts. These are generalised in the following manner.

Proposition A.2.9

(1)   Let (X, τ) be a topological space. The following conditions on a subcollection $\mathcal{B} \subset \tau$ are equivalent:

  (i) $U \in \tau \Rightarrow \exists \{B_i : i \in I\} \subset \mathcal{B}$ such that $U = \cup_{i \in I} B_i$;
  (ii) $x \in U, U$ open $\Rightarrow \exists B \in \mathcal{B}$ such that $x \in B \subset U$.

A collection $\mathcal{B}$ satisfying these equivalent conditions is called a **base** for the topology τ.

(2)   A family $\mathcal{B}$ is a base for some topology τ on X if and only if $\mathcal{B}$ satisfies the two following conditions:

  (a) $\mathcal{B}$ covers X, meaning that $X = \cup_{B \in \mathcal{B}} B$; and
  (b) $B_1, B_2 \in \mathcal{B}, x \in B_1 \cap B_2 \Rightarrow \exists B \in \mathcal{B}$ such that $x \in B \subset (B_1 \cap B_2)$.

The elementary proof of this proposition is left as an exercise for the reader. It should be fairly clear that if $\mathcal{B}$ is a base for a topology τ on X, and if τ' is any topology such that $\mathcal{B} \subset \tau'$, then, necessarily $\tau \subset \tau'$. Thus, if $\mathcal{B}$ is a base for a topology τ, then τ is the 'smallest topology' with respect to which all the members of $\mathcal{B}$ are open. However, as condition (ii) of Proposition A.2.9 shows, not any collection of sets can be a base for some topology. This state of affairs is partially remedied in the following proposition.

Proposition A.2.10

(a)   Let X be a set and let $\mathcal{S}$ be an arbitrary family of subsets of X. Then there exists a smallest topology $\tau(\mathcal{S})$ on X such that $\mathcal{S} \subset \tau(\mathcal{S})$; we shall refer to $\tau(\mathcal{S})$ as the topology generated by $\mathcal{S}$.
(b)   Let $X, \mathcal{S}, \tau(\mathcal{S})$ be as in (a) above. Let $\mathcal{B} = \{X\} \cup \{B : \exists n \in \mathbb{N},$ and $S_1, S_2, \dots, S_n \in \mathcal{S}$ such that $B = \cap_{i=1}^n S_i\}$. Then $\mathcal{B}$ is a base for the topology τ; in particular, a typical element of $\tau(\mathcal{S})$ is expressible as an arbitrary union of finite intersections of members of $\mathcal{S}$.

If (X, τ) is a topological space, a family $\mathcal{S}$ is said to be a **sub-base** for the topology τ if it is the case that $\tau = \tau(\mathcal{S})$.

Proof: For (a), we may simply define

$$\tau(\mathcal{S}) = \cap \{\tau' : \tau' \text{ is a topology and } \mathcal{S} \subset \tau'\},$$

and note that this does the job.

As for (b), it is clear that the family $\mathcal{B}$ as defined in (b) covers X and is closed under finite intersections. Consequently, if we define $\tau = \{\cup_{B \in \mathcal{B}_0} B : \mathcal{B}_0 \subset \mathcal{B}\}$, it may be verified that τ is a topology on X for which $\mathcal{B}$ is a base; on the other hand, it is clear from the construction (and the definition of a topology) that if τ' is any topology which contains $\mathcal{S}$, then τ' must necessarily contain τ, and, we find that $\tau = \tau(\mathcal{S})$.   □

The usefulness of sub-bases may be seen as follows: very often, in wanting to define a topology, we will find that it is natural to require that sets belonging to a certain class $\mathcal{S}$ should be open; then the topology we seek is any topology which is at least large as $\tau(\mathcal{S})$. We will find that this minimal topology is quite a good one to work with since we know by the last proposition precisely what the open sets in this topology look like. In order to make all this precise as well as for other reasons, we need to discuss the notion of continuity in the context of topological spaces.

Definition A.2.11   *A function $f : X \to Y$ between topological spaces is said to be:*

(a)   *continuous at the point $x \in X$ if $f^{-1}(U)$ is an open neighbourhood of x in the topological space X whenever U is an open neighbourhood of the point $f(x)$ in the topological space Y;*

(b)   **continuous** *if it is continuous at each $x \in X$, or equivalently, if $f^{-1}(U)$ is an open set in X whenever U is an open set in Y.*

The proof of the following elementary proposition is left as an exercise for the reader.

Proposition A.2.12   *Let $f : X \to Y$ be a map between topological spaces.*

(1)   *If $x \in X$, then f is continuous at x if and only if $\{f(x_i) : i \in I\}$ is a net converging to $f(x)$ in Y whenever $\{x_i : i \in I\}$ is a net converging to x in X.*

(2)   *The following conditions on f are equivalent:*

    (i)   *f is continuous;*

    (ii)   *$f^{-1}(F)$ is a closed subset of X whenever F is a closed subset of Y;*

    (iii)   *$\{f(x_i) : i \in I\}$ is a net converging to $f(x)$ in Y whenever $\{x_i : i \in I\}$ is a net converging to x in X;*

    (iv)   *$f^{-1}(B)$ is open in X whenever B belongs to some base for the topology on Y;*

    (v)   *$f^{-1}(S)$ is open in X whenever S belongs to some sub-base for the topology on Y.*

(3)   *The composition of continuous maps is continuous, i.e., if $f : X \to Y$ and $g : Y \to Z$ are continuous maps of topological spaces, then $g \circ f : X \to Z$ is continuous.*

We are now ready to illustrate what we meant in our earlier discussion on the usefulness of sub-bases. Typically, we have the following situation in mind: suppose X is just some set, $\{Y_i : i \in I\}$ is some family of topological spaces and suppose that we have maps $f_i : X \to Y_i, \forall i \in I$. We would want to topologise X in such a way that each of the maps f_i is continuous. This, by itself, is not difficult since if X is equipped with the discrete topology, any map from X into any topological space would be continuous; but if we want to topologise X in an efficient as well as natural manner with respect to the requirement that each f_i is continuous, then the method of sub-bases tells us what to do. Let us make all this explicit.

Proposition A.2.13   *Suppose X is a set and $\{f_i : X \to X_i | i \in I\}$ is a family of functions and suppose τ_i is a topology on X_i for each $i \in I$. Let S_i be an arbitrary sub-base for the topology τ_i. Define $\tau = \tau(S)$, where*

$$S = \{f_i^{-1}(V_i) : V_i \in S_i, i \in I\}.$$

Then,

(a)   *f_i is continuous as a mapping from the topological space (X, τ) into the topological space (X_i, τ_i) for each $i \in I$;*
(b)   *the topology τ is the smallest topology on X for which (a) above is valid and consequently, this topology is independent of the choice of the sub-bases $S_i, i \in I$ and depends only upon the data $\{f_i, \tau_i, i \in I\}$.*

This topology τ on X is called the **weak topology** *induced by the family $\{f_i : i \in I\}$ of functions, and we shall write $\tau = \tau(\{f_i : i \in I\})$.*

The proposition is an immediate consequence of Proposition A.2.12(2)(ii) and Proposition A.2.10.

Ex. A.2.14   Let X, X_i, S_i, τ_i, f_i and $\tau = \tau(\{f_i : i \in I\})$ be as in Proposition A.2.13.

(a)   Suppose Z is a topological space and $g : Z \to X$ is a function. Then, show that g is continuous if and only if $f_i \circ g$ is continuous for each $i \in I$.
(b)   Show that the family $\mathcal{B} = \{\cap_{j=1}^{n} f_{i_j}^{-1}(V_{i_j}) : i_1, \ldots, i_n \in I, n \in \mathbb{N}, V_{i_j} \in \tau_{i_j} \forall j\}$ is a base for the topology τ.
(c)   Show that a net $\{x_\lambda : \lambda \in \Lambda\}$ converges to x in (X, τ) if and only if the net $\{f_i(x_\lambda) : \lambda \in \Lambda\}$ converges to $f_i(x)$ in (X_i, τ_i), for each $i \in I$.

As in a metric space any subspace (= subset) of a topological space acquires a natural structure of a topological space in the manner indicated in the following exercise.

Ex. A.2.15 Let (Y, τ) be a topological space, and let $X \subset Y$ be a subset. Let $i_{X \to Y} : X \to Y$ denote the inclusion map. Then the **subspace topology** on X (or the topology on X induced by the topology τ on Y) is, by definition, the weak topology $\tau(\{i_{X \to Y}\})$. This topology will be denoted by $\tau|_X$.

(a) Show that $\tau|_X = \{U \cap X : U \in \tau\}$, or equivalently, that a subset $F \subset X$ is closed in $(X, \tau|_X)$ if and only if there exists a closed set F_1 in (Y, τ) such that $F = F_1 \cap X$.

(b) Show that if Z is some topological space and if $f : Z \to X$ is a function, then f is continuous when regarded as a map into the topological space $(X, \tau|_X)$ if and only if it is continuous when regarded as a map into the topological space (Y, τ).

One of the most important special cases of the weak topology is the **product** of topological spaces. Suppose $\{(X_i, \tau_i) : i \in I\}$ is an arbitrary family of topological spaces. Let $X = \prod_{i \in I} X_i$ denote their Cartesian product. Then the **product topology** is, by definition, the weak topology on X induced by the family $\{\pi_i : X \to X_i | i \in I\}$, where π_i denotes, for each $i \in I$, the natural 'projection' of X onto X_i. We shall denote this product topology by $\prod_{i \in I} \tau_i$. Note that if $\mathcal{B}_i$ is a base (resp., $\mathcal{S}_i$ is a sub-base) for the topology τ_i, then a base (resp. sub-base) for the product topology is given by the family $\mathcal{B}$ (resp. $\mathcal{S}$), where a typical element of $\mathcal{B}$ (resp. $\mathcal{S}$) has the form

$$U = \{x \in X : \pi_i(x) \in U_i \ \forall i \in I_0\},$$

where I_0 is an arbitrary finite (resp. singleton) subset of I and $U_i \in \mathcal{B}_i$ (resp. $U_i \in \mathcal{S}_i$) for each $i \in I_0$. Thus, a typical basic open set is prescribed by constraining some *finitely* many coordinates to lie in specified basic open sets in the appropriate spaces.

Note that if Z is any set, then functions $F : Z \to X$ are in a 1-1 correspondence with families $\{f_i : Z \to X_i | i \in I\}$ of functions, where $f_i = \pi_i \circ f$; it follows from Ex. A.2.14 that if Z is a topological space, then the function f is continuous precisely when each f_i is continuous. To make sure that you have really understood the definition of the product topology, you are urged to work out the following exercise.

Ex. A.2.16 If (X, τ) is a topological space and if I is any set (let X^I denote the space of functions $f : I \to X$):

(a)   Show that X^I may be identified with the product $\prod_{i \in I} X_i$, where $X_i = X \ \forall i \in I$.

(b)   Let X^I be equipped with product topology; fix $x_0 \in X$ and show that the set $D = \{ f \in X^I : f(i) = x_0$ for all but a finite number of i's $\}$ is dense in X, meaning that if U is any open set in X, then $D \cap U \neq \emptyset$.

(c)   If Λ is a directed set, show that a net $\{ f_\lambda : \lambda \in \Lambda \}$ in X^I converges to a point $f \in X^I$ if and only if the net $\{ f_\lambda(i) : \lambda \in \Lambda \}$ converges to $f(i)$ for each $i \in I$. (In other words, the product topology on X^I is nothing but the topology of 'pointwise convergence'.)

We conclude this section with a brief discussion of 'homeomorphisms'.

Definition A.2.17   *Two topological spaces X and Y are said to be* **homeomorphic** *if there exists continuous functions $f : X \rightarrow Y, g : Y \rightarrow X$ such that $f \circ g = id_Y$ and $g \circ f = id_X$. A* **homeomorphism** *is a map f as above, i.e. a continuous bijection between two topological spaces whose (set-theoretic) inverse is also continuous.*

The reader should observe that requiring a function $f : X \rightarrow Y$ is a homeomorphism is more than just requiring that f is 1-1, onto and continuous. If only so much is required of the function, then the inverse f^{-1} may fail to be continuous. An example of this phenomenon is provided by the function $f : [0, 1) \rightarrow \mathbb{T}$ defined by $f(t) = \exp(2\pi i t)$.

   The proof of the following proposition is elementary and is left as an exercise to the reader.

Proposition A.2.18   *Suppose f is a continuous bijective map of a topological space X onto a space Y, then the following conditions are equivalent:*

(i)   *f is a homeomorphism;*

(ii)   *f is an* **open map**,*i.e., if U is an open set in X, then $f(U)$ is an open set in Y;*

(iii) *f is a* **closed map**, *i.e., if F is a closed set in X, then $f(F)$ is a closed set in Y.*

A.3   Compactness

This section is devoted to a quick review of the theory of compact spaces. For the uninitiated reader, the best way to get a feeling for compactness is by understanding this notion for subsets of the real line. The features of compact subsets of $\mathbb{R}$ (and, more generally, of any $\mathbb{R}^n$) are summarised in the following result.

Theorem A.3.1 *The following conditions on a subset $K \subset \mathbb{R}^n$ are equivalent:*

(i)   *K is closed and bounded;*

(ii)   *every sequence in K has a subsequence which converges to some point in K;*

(iii) *every **open cover** of K has a finite sub-cover, i.e., if $\{U_i : i \in I\}$ is any collection of open sets such that $K \subset \cup_{i \in I} U_i$, then there exists a finite subcollection $\{U_{i_j} : 1 \leq j \leq n\}$ which still covers K (meaning that $K \subset \cup_{j=1}^n U_{i_j}$).*

(iv) *Suppose $\{F_i : i \in I\}$ is any family of closed subsets which has the **finite intersection property** with respect to K, i.e., $\cap_{i \in I_0} F_i \cap K \neq \emptyset$ for any finite subset $I_0 \subset I$, then it is necessarily the case that $\cap_{i \in I} F_i \cap K \neq \emptyset$.*

*A subset $K \subset \mathbb{R}$ which satisfies the equivalent conditions above is said to be **compact**.*   □

We will not prove this here, since we shall be proving more general statements.

To start with, note that the conditions (iii) and (iv) of Theorem A.3.1 make sense for any subset K of a topological space X, and are easily seen to be equivalent (by considering the equation $F_i = X - U_i$); also, while conditions (i) and (ii) make sense in any topological space, they may not be very strong in general. What we shall see is that conditions (ii) - (iv) are equivalent for any subset of a metric space, and these are equivalent to a stronger version of (i). We begin with the appropriate definition.

Definition A.3.2 *A subset K of a topological space is said to be **compact** if it satisfies the equivalent conditions (iii) and (iv) of Theorem A.3.1.*

'Compactness' is an *intrinsic property*, as asserted in the following exercise.

Ex. A.3.3 Suppose K is a subset of a topological space (X, τ). Then show that K is compact when regarded as a subset of the topological space (X, τ) if and only if K is compact when regarded as a subset of the topological space $(K, \tau|_K)$; see Ex. A.2.15.

Proposition A.3.4

(a) *Let $\mathcal{B}$ be any base for the topology underlying a topological space X. Then, a subset $K \subset X$ is compact if and only if any cover of K by open sets all of which belong to the base $\mathcal{B}$ admits a finite sub-cover.*

(b) *A closed subset of a compact set is compact.*

Proof:

(a)   Clearly, every 'basic open cover' of a compact set admits a finite subcover. Conversely, suppose that $\{U_i : i \in I\}$ is an open cover of a set K and that every cover of K by members of $\mathcal{B}$ admits a finite subcover. For each $i \in I$, find a sub-family $\mathcal{B}_i \subset \mathcal{B}$ such that $U_i = \cup_{B \in \mathcal{B}_i} B$. Then $\{B : B \in \mathcal{B}_i, i \in I\}$ is a cover of K by members of $\mathcal{B}$, and so there exist $B_1, \ldots, B_n$ in this family such that $K \subset \cup_{i=1}^{n} B_i$. For each $j = 1, \ldots, n$ by construction, we can find $i_j \in I$ such that $B_j \subset U_{i_j}$. It follows that $K \subset \cup_{j=1}^{n} U_{i_j}$.

(b)   Suppose $C \subset K \subset X$, where K is a compact subset of X and C is a subset of K which is closed in the subspace topology of K. Thus, there exists a closed set $F \subset X$ such that $C = F \cap K$. Now, suppose $\{U_i : i \in I\}$ is an open cover of C. Then $\{U_i : i \in I\} \cup \{X - F\}$ is an open cover of K. So there exists a finite subfamily $I_0 \subset I$ such that $K \subset \cup_{i \in I_0} U_i \cup (X - F)$, which clearly implies that $C \subset \cup_{i \in I_0} U_i$.   □

Corollary A.3.5   *Let K be a compact subset of a metric space X. Then, for any $\epsilon > 0$, there exists a finite subset $N_\epsilon \subset K$ such that $K \subset \cup_{x \in N_\epsilon} B(x, \epsilon)$, (where, of course, the symbol $B(x, \epsilon) = \{y \in X : d(y, x) < \epsilon\}$ denotes the open ball with centre X and radius ϵ).*

Any set N_ϵ as above is called an ϵ-net for K.

Proof: The family of all open balls with (arbitrary centres and) radii bounded by $\frac{\epsilon}{2}$ clearly constitutes a base for the topology of the metric space X. (Verify this!) Hence, by the preceding proposition, we can find $y_1, \ldots, y_n \in X$ such that $K \subset \cup_{i=1}^{n} B(y_i, \frac{\epsilon}{2})$. We may assume, without loss of generality, that K is not contained in any proper sub-family of these n open balls, i.e., we may assume that there exists $x_i \in K \cap B(y_i, \frac{\epsilon}{2})$, $\forall i$. Then, clearly, $K \subset \cup_{i=1}^{n} B(x_i, \epsilon)$ and so $N_\epsilon = \{x_1, \ldots, x_n\}$ is an ϵ-net for K.   □

Definition A.3.6   *A subset A of a metric space X is said to be **totally bounded** if, for each $\epsilon > 0$, there exists a finite ϵ-net for A.*

Thus, compact subsets of metric spaces are totally bounded. The next proposition provides an alternative criterion for total boundedness.

Proposition A.3.7   *The following conditions on a subset A of a metric space X are equivalent:*

(i)   *A is totally bounded;*
(ii)   *every sequence in A has a Cauchy subsequence.*

Proof: $(i) \Rightarrow (ii)$: Suppose $S_0 = \{x_k\}_{k=1}^{\infty}$ is a sequence in A. We assert that there exist (a) a sequence $\{B_n = B(z_n, 2^{-n})\}$ of open balls (with shrinking radii) in X; and (b) sequences $S_n = \{x_k^{(n)}\}_{k=1}^{\infty}, n = 1, 2, \ldots$ with the property that $S_n \subset B_n$ and S_n is a subsequence of S_{n-1}, for all $n \geq 1$.

We construct the B_n's and the S_n's inductively. To start with, pick a finite $\frac{1}{2}$-net $N_{\frac{1}{2}}$ for A; clearly, there must be some $z_1 \in N_{\frac{1}{2}}$ with the property that $x_k \in B(z_1, \frac{1}{2})$ for infinitely-many values of k. Define $B_1 = B(z_1, \frac{1}{2})$ and choose S_1 to be any subsequence $\{x_k^{(1)}\}_{k=1}^{\infty}$ of S_0 with the property that $x_k^{(1)} \in B_1 \, \forall k$.

Suppose now that open balls $B_j = B(z_j, 2^{-j})$ and sequences $S_j, 1 \leq j \leq n$ have been chosen so that S_j is a subsequence of S_{j-1} which lies in $B_j \, \forall 1 \leq j \leq n$.

Now, let $N_{2^{-(n+1)}}$ be a finite $2^{-(n+1)}$-net for A. As before, we may argue that there must exist some $z_{n+1} \in N_{2^{-(n+1)}}$ with the property that if $B_{n+1} = B(z_{n+1}, 2^{-(n+1)})$, then $x_k^{(n)} \in B_{n+1}$ for infinitely-many values of k. We choose S_{n+1} to be a subsequence of S_n which lies entirely in B_{n+1}.

Thus, by induction, we can conclude the existence of sequences of open balls B_n and sequences S_n with the asserted properties. Now, define $y_n = x_n^{(n)}$ and note that (i) $\{y_n\}$ is a subsequence of the initial sequence S_0, and (ii) if $n, m \geq k$, then $y_n, y_m \in S_k \subset B_k$ and hence $d(y_n, y_m) < 2^{1-k}$; this is the desired Cauchy subsequence.

$(ii) \Rightarrow (i)$: Suppose A is not totally bounded. Then there exists some $\epsilon > 0$ such that A admits no finite ϵ-net. This means that given any finite subset $F \subset A$, there exists some $a \in A$ such that $d(a, x) \geq \epsilon \, \forall \, x \in F$. Pick $x_1 \in A$ (this is possible since we may take $F = \emptyset$ in the previous sentence), then pick $x_2 \in A$ such that $d(x_1, x_2) \geq \epsilon$ (this is possible by setting $F = \{x_1\}$ in the previous sentence), then (set $F = \{x_1, x_2\}$ in the previous sentence and) pick $x_3 \in A$ such that $d(x_3, x_i) \geq \epsilon, i = 1, 2$, and so on. This yields a sequence $\{x_n\}_{n=1}^{\infty}$ in A such that $d(x_n, x_m) \geq \epsilon \, \forall \, n \neq m$. This sequence clearly has no Cauchy subsequence, thereby contradicting the assumption (ii). The proof is complete. $\square$

Corollary A.3.8   *If A is a totally bounded set in a metric space, then so also are its closure $\overline{A}$ and any subset of A.*

Proof: Exercise.

Remark A.3.9   The argument to prove $(i) \Rightarrow (ii)$ in the above theorem has a very useful component which we wish to single out. Starting with a sequence $S = \{x_k\}$, we proceeded by some method (which method was dictated by the specific purpose on hand is not relevant here) to construct sequences $S_n = \{x_k^{(n)}\}$ with the property that each S_{n+1} was a subsequence of S_n (and with some additional desirable properties); we then considered the sequence $\{x_n^{(n)}\}_{n=1}^{\infty}$, which is a subsequence of each S_n. This process is sometimes referred to as the **diagonal argument**.

Ex. A.3.10

(1)   Show that any totally bounded set in a metric space is bounded, meaning that it is a subset of some open ball.
(2)   Show that a subset of $\mathbb{R}^n$ is bounded if and only if it is totally bounded. (*Hint*: by Corollary A.3.8 and (1) above, it is enough to establish the total-boundedness of the set given by $K = \{x = (x_1, \ldots, x_n) : |x_i| \leq N \; \forall i\}$, and given ϵ, pick k such that the diameter of an n-cube of side $\frac{2}{k}$ is smaller than ϵ, and then consider points of the form $\left(\frac{i_1}{k}, \frac{i_2}{k}, \ldots, \frac{i_n}{k} \right)$ in order to find an ϵ-net.)

We are almost ready to prove the announced equivalence of the various conditions for compactness, but first, we need a technical result which we look into before proceeding to the desired assertion.

Lemma A.3.11   *The following conditions on a metric space X are equivalent:*

(i)   *X is **separable**, i.e., there exists a countable set D which is dense in X (meaning that $X = \overline{D}$);*
(ii)   *X satisfies the **second axiom of countability**, meaning that there is a countable base for the topology of X.*

Proof: $(i) \Rightarrow (ii)$: Let $\mathcal{B} = \{B(x, r) : x \in D, 0 < r \in \mathbb{Q}\}$ where D is some countable dense set in X and $\mathbb{Q}$ denotes the (countable) set of rational numbers. We assert that this is a base for the topology on X, for, if U is an open set and if $y \in U$, we may find $s > 0$ such that $B(y, s) \subset U$. Pick $x \in D \cap B(y, \frac{s}{3})$ and pick a positive rational number r such that $d(x, y) < r < \frac{s}{2}$ and note that $y \in B(x, r) \subset U$. Thus, for each $y \in U$, we have found a ball $B_y \in \mathcal{B}$ such that $y \in B_y \subset U$. Hence $U = \cup_{y \in U} B_y$.

$(ii) \Rightarrow (i)$: If $\mathcal{B} = \{B_n\}_{n=1}^{\infty}$ is a countable base for the topology of X (where we may assume that each B_n is a non-empty set without loss of generality), pick a point $x_n \in B_n$ for each n and verify that $D = \{x_n\}_{n=1}^{\infty}$ is indeed a countable dense set in X. $\square$

Theorem A.3.12   *The following conditions on a subset K of a metric space X are equivalent:*

(i)   *K is compact;*
(ii)   *K is complete and totally bounded;*

(iii)   *Every sequence in K has a subsequence which converges to some point of*
K.

Proof: $(i) \Rightarrow (ii)$: We have already seen that compactness implies total boundedness. Suppose now that $\{x_n\}$ is a Cauchy sequence in K. Let F_n be the closure
of the set $\{x_m : m \geq n\}$ for all $n = 1, 2, \ldots$. Then, $\{F_n\}_{n=1}^\infty$ is clearly a
decreasing sequence of closed sets in X; further, the Cauchy criterion implies
that diam $F_n \to 0$ as $n \to \infty$. By invoking the finite intersection property, we see
that there must exist a point $x \in \cap_{n=1}^\infty F_n \cap K$. Since the diameters of the F_n's
shrink to 0, we may conclude that (such an x is unique) $x_n \to x$ as $n \to \infty$.

$(ii) \Leftrightarrow (iii)$: This is an immediate consequence of Proposition A.3.7.

$(ii) \Rightarrow (i)$: For each $n = 1, 2, \ldots$, let $N_{\frac{1}{n}}$ be a $\frac{1}{n}$-net for K. It should be clear
that $D = \cup_{n=1}^\infty N_{\frac{1}{n}}$ is a countable dense set in K. Consequently, K is separable.
Hence, by Lemma A.3.11, K admits a countable base $\mathcal{B}$.

In order to check that K is compact in X, it is enough, by Ex. A.3.3, to check
that K is compact in itself. Hence, by Proposition A.3.4, it is enough to check
that any countable open cover of X has a finite sub-cover by the reasoning that
goes in to establish the equivalence of conditions (iii) and (iv) of Theorem A.3.1.
We have, therefore, to show that if $\{F_n\}_{n=1}^\infty$ is a sequence of closed sets in K
such that $\cap_{n=1}^N F_n \neq \emptyset$ for each $N = 1, 2, \ldots$. Then, necessarily, $\cap_{n=1}^\infty F_n \neq \emptyset$.
For this, pick a point $x_N \in \cap_{n=1}^N F_n$ for each N; appeal to the hypothesis to find
a subsequence $\{y_n\}$ of $\{x_n\}$ such that $y_n \to x \in K$ for some point $x \in K$. Now
notice that $\{y_m\}_{m=N}^\infty$ is a sequence in $\cap_{n=1}^N F_n$ which converges to x; conclude
that $x \in F_n \forall n$. Hence $\cap_{n=1}^\infty F_n \neq \emptyset$. $\square$

In view of Ex. A.3.10(2), we see that Theorem A.3.12 does indeed generalise
Theorem A.3.1.

We now wish to discuss compactness in general topological spaces. We begin
with some elementary results.

Proposition A.3.13

(a)   *Suppose $f : X \to Y$ is a continuous map of topological spaces. If K is a
compact subset of X, then $f(K)$ is a compact subset of Y; in other words,
a continuous image of a compact set is compact.*

(b)   *If $f : K \to \mathbb{R}$ is continuous and if K is a compact set in X, then (i) $f(K)$
is bounded, and (ii) there exist points $x, y \in K$ such that $f(x) \leq f(z) \leq
f(y) \ \forall \ z \in K$; in other words, a continuous real-valued function on a
compact set is bounded and attains its bounds.*

Proof:

(a)   If $\{U_i : i \in I\}$ is an open cover of $f(K)$, then $\{f^{-1}(U_i) : i \in I\}$ is an
open cover of K; if $\{f^{-1}(U_i) : i \in I_0\}$ is a finite sub-cover of K, then
$\{U_i : i \in I_0\}$ is a finite sub-cover of $f(K)$.

(b)   This follows from (a) since compact subsets of $\mathbb{R}$ are closed and bounded and hence contain their supremum and infimum.   □

The following result is considerably stronger than Proposition A.3.4(a).

Theorem A.3.14   (Alexander sub-base theorem) *Let S be a sub-base for the topology τ underlying a topological space X. Then a subspace $K \subset X$ is compact if and only if any open cover of K by a sub-family of S admits a finite sub-cover.*

Proof: Suppose S is a sub-base for the topology of X and suppose that any open cover of K by a sub-family of S admits a finite sub-cover. Let B be the base generated by S, i.e., a typical element of B is a finite intersection of members of S. In view of Proposition A.3.4(a), the theorem will be proved once we establish that any cover of X by members of B admits a finite sub-cover.

Assume this is false. Thus, suppose $\mathcal{U}_0 = \{B_i : i \in I_0\} \subset B$ is an open cover of X which does not admit a finite subcover. Let $\mathcal{P}$ denote the set of all subfamilies $\mathcal{U} \subset \mathcal{U}_0$ with the property that no finite sub-family of $\mathcal{U}$ covers X. (Notice that $\mathcal{U}$ is an open cover of X since $\mathcal{U} \subset \mathcal{U}_0$.) It should be obvious that $\mathcal{P}$ is partially ordered by inclusion, and that $\mathcal{P} \neq \emptyset$ since $\mathcal{U} \in \mathcal{P}$. Suppose $\{\mathcal{U}_i : i \in I\}$ is a totally ordered subset of $\mathcal{P}$. We assert that then $\mathcal{U} = \cup_{i \in I} \mathcal{U}_i \in \mathcal{P}$. (Reason: Clearly $\mathcal{U}_0 \subset \cup_{i \in I} \mathcal{U}_i \subset B$; further, suppose $\{B_1, \ldots, B_n\} \subset \cup_{i \in I} \mathcal{U}_i$; the assumption of the family $\{\mathcal{U}_i\}$ being totally ordered then shows that in fact there must exist some $i \in I$ such that $\{B_1, \ldots, B_n\} \subset \mathcal{U}_i$. By definition of $\mathcal{P}$, it cannot be the case that $\{B_1, \ldots, B_n\}$ covers X. Thus, we have shown that no finite sub-family of $\mathcal{U}$ covers X. Hence, Zorn's lemma is applicable to the partially ordered set $\mathcal{P}$.

Thus, if the theorem were false, it would be possible to find a family $\mathcal{U} \subset B$ which is an open cover of X, and further it has the following properties: (i) no finite sub-family of $\mathcal{U}$ covers X; and (ii) $\mathcal{U}$ is a maximal element of $\mathcal{P}$, which means that whenever $B \in B - \mathcal{U}$, there exists a finite subfamily $\mathcal{U}_B$ of $\mathcal{U}$ such that $\mathcal{U}_B \cup \{B\}$ is a (finite) cover of X.

Now, by definition of B, each element B of $\mathcal{U}$ has the form $B = S_1 \cap S_2 \cdots \cap S_n$ for some $S_1, \ldots, S_n \in S$. Assume for the moment that none of the S_i's belongs to $\mathcal{U}$. Then property (ii) of $\mathcal{U}$ (in the previous paragraph) implies that for each $1 \leq i \leq n$, we can find a finite sub-family $\mathcal{U}_i \subset \mathcal{U}$ such that $\mathcal{U}_i \cup \{S_i\}$ is a finite open cover of X. Since this is true for all i, this implies that $\cup_{i=1}^{n} \mathcal{U}_i \cup \{B = \cap_{i=1}^{n} S_i\}$ is a finite open cover of X, but since $B \in \mathcal{U}$, we have produced a finite sub-cover of X from $\mathcal{U}$, which contradicts the defining property (i) (in the previous paragraph) of $\mathcal{U}$. Hence at least one of the S_i's must belong to $\mathcal{U}$.

Thus, we have shown that if $\mathcal{U}$ is as above, then each $B \in \mathcal{U}$ is of the form $B = \cap_{i=1}^{n} S_i$, where $S_i \in S\ \forall i$ and further, there exists at least one i_0 such that

$S_{i_0} \in \mathcal{U}$. The passage $B \mapsto S_{i_0}$ yields a mapping $\mathcal{U} \ni B \mapsto S(B) \in \mathcal{U} \cap S$ with the property that $B \subset S(B)$ for all $B \in \mathcal{U}$. Since $\mathcal{U}$ is an open cover of X by definition, we find that $\mathcal{U} \cap S$ is an open cover of X by members of S, which admits a finite sub-cover by hypothesis.

The contradiction we have reached is a consequence of our assuming that there exists an open cover $\mathcal{U}_0 \subset B$ which does not admit a finite sub-cover. The proof of the theorem is finally complete.   □

We are now ready to prove the important result, due to Tychonoff, which asserts that if a family of topological spaces is compact, then so is their topological product.

Theorem A.3.15   **(Tychonoff's theorem)** *Suppose* $\{(X_i, \tau_i) : i \in I\}$ *is a family of non-empty topological spaces. Let* (X, τ) *denote the product space* $X = \prod_{i \in I} X_i$ *equipped with the product topology. Then* X *is compact if and only if each* X_i *is compact.*

Proof: Since $X_i = \pi_i(X)$, where π_i denotes the (continuous) projection onto the ith co-ordinate, it follows from Proposition A.3.13(a) that if X is compact, so is each X_i.

Suppose conversely that each X_i is compact. For a subset $A \subset X_i$, let $A^i = \pi_i^{-1}(A)$. By the definition of the product topology, $S = \{U^i : U \in \tau_i\}$ is a sub-base for the product topology τ. Thus, we need to prove the following: if J is any set, if $J \ni j \mapsto i(j) \in I$ is a map, if $A(i(j))$ is a closed set in $X_{i(j)}$ for each $j \in J$, and if $\cap_{j \in F} A(i(j))^{i(j)} \neq \emptyset$ for every finite subset $F \subset J$, then it is necessarily the case that $\cap_{j \in J} A(i(j))^{i(j)} \neq \emptyset$.

Let $I_1 = \{i(j) : j \in J\}$ denote the range of the mapping $j \mapsto i(j)$. For each fixed $i \in I_1$, let $J_i = \{j \in J : i(j) = i\}$. Observe that $\{A(i(j)) : j \in J_i\}$ is a family of closed sets in X_i. We assert that $\cap_{j \in F} A(i(j)) \neq \emptyset$ for every finite subset $F \subset J_i$. (Reason: if this were empty, then, $\cap_{j \in F} A(i(j))^{i(j)} = (\cap_{j \in F} A(i(j)))^i$ would have to be empty, which would contradict the hypothesis.) We conclude from the compactness of X_i that $\cap_{j \in J_i} A(i(j)) \neq \emptyset$. Let x_i be any point from this intersection. Thus $x_i \in A(i(j))$ whenever $j \in J$ and $i(j) = i$. For $i \in I - I_1$, pick an arbitrary point $x_i \in X_i$. It is now readily verified that if $x \in X$ is the point such that $\pi_i(x) = x_i \; \forall i \in I$, then $x \in A(i(j))^{i(j)} \; \forall j \in J$; the proof of the theorem is complete.   □

Among topological spaces there is an important subclass of spaces which exhibit some pleasing features (in that certain pathological situations cannot occur. We briefly discuss these spaces now.

Definition A.3.16   *A topological space is called a* **Hausdorff space** *if, whenever x and y are two distinct points in X, there exist open sets U, V in X such that $x \in U, y \in V$ and $U \cap V = \emptyset$. Thus, any two distinct points* *can be 'separated' (by a pair of disjoint open sets). (For obvious reasons, the preceding requirement is sometimes also referred to as the Hausdorff separation axiom.)*

Ex. A.3.17

(1)   Show that any metric space is a Hausdorff space.

(2)   Show that if $\{f_i : X \to X_i : i \in I\}$ is a family of functions from a set X into topological spaces (X_i, τ_i), and if each (X_i, τ_i) is a Hausdorff space, then the weak topology $\tau(\{f_i : i \in I\})$ on X (which is induced by this family of maps) is a Hausdorff topology if and only if the f_i's are separate points - meaning that whenever x and y are distinct points in X, then there exists an index $i \in I$ such that $f_i(x) \neq f_i(y)$.

(3)   Show that if $\{(X_i, \tau_i) : i \in I\}$ is a family of topological spaces, then the topological product space $(\prod_{i \in I} X_i, \prod_{i \in I} \tau_i)$ is a Hausdorff space if and only if each (X_i, τ_i) is a Hausdorff space.

(4)   Show that a subspace of a Hausdorff space is also a Hausdorff space (with respect to the subspace topology).

(5)   Show that Exercise (4) as well as the 'only if' assertion of Exercise (3) above are consequences of Exercise (2).

We list some elementary properties of Hausdorff spaces in the following proposition.

Proposition A.3.18

(a)   *If (X, τ) is a Hausdorff space, then every finite subset of X is closed.*

(b)   *A topological space is a Hausdorff space if and only if the 'limits of convergent nets are unique', i.e., if and only if, whenever $\{x_i : i \in I\}$ is a net in X, and if the net converges to both $x \in X$ and $y \in X$, then $x = y$.*

(c)   *Suppose K is a compact subset of a Hausdorff space X and suppose $y \notin K$; then there exist open sets U, V in X such that $K \subset U, y \in V$ and $U \cap V = \emptyset$; in particular, a compact subset of a Hausdorff space is closed.*

(d)   *If X is a compact Hausdorff space, and if C and K are closed subsets of X which are disjoint, i.e., if $C \cap K = \emptyset$, then there exist a pair of disjoint open sets U, V in X such that $K \subset U$ and $C \subset V$.*

Proof:

(a)   Since finite unions of closed sets are closed, it is enough to prove that $\{x\}$ is closed, for each $x \in X$; the Hausdorff separation axiom clearly implies that $X - \{x\}$ is open.

(b)   Suppose X is Hausdorff, suppose a net $\{x_i : i \in I\}$ converges to $x \in X$ and suppose $x \neq y \in X$. Pick open sets U, V as in Definition A.3.16. By definition of convergence of a net, we can find $i_0 \in I$ such that $x_i \in U$ $\forall i \geq i_0$. It follows then that $x_i \notin V$ $\forall i \geq i_0$, and hence the net $\{x_i\}$ clearly does not converge to y.

Conversely, suppose X is not a Hausdorff space; then there exists a pair x, y of distinct points which cannot be separated. Let $\mathcal{N}(x)$ (resp., $\mathcal{N}(y)$) denote the directed set of open neighbourhoods of the point x (resp., y) (see Example A.2.4(3)). Let $I = \mathcal{N}(x) \times \mathcal{N}(y)$ be the directed set obtained from the Cartesian product as in Example A.2.4(4). By the assumption that x and y cannot be separated, we can find a point $x_i \in U \cap V$ for each $i = (U, V) \in I$. It is fairly easily seen that the net $\{x_i : i \in I\}$ simultaneously converges to both x and y.

(c)   Suppose K is a compact subset of a Hausdorff space X. Fix $y \notin K$. Then, for each $x \in K$, find open sets U_x, V_x so that $x \in U_x$, $y \in V_x$ and $U_x \cap V_x = \emptyset$. Now the family $\{U_x : x \in K\}$ is an open cover of the compact set K, and we can hence find $x_1, \ldots, x_n \in K$ such that $K \subset U = \bigcup_{i=1}^n U_{x_i}$. Conclude then that if $V = \bigcap_{i=1}^n V_{x_i}$, V is an open neighbourhood of y such that $U \cap V = \emptyset$. The proof of (c) is complete.

(d)   Since closed subsets of compact spaces are compact, we see that C and K are compact. Hence, by (c) above, we may, for each $y \in C$, find a pair U_y, V_y of disjoint open subsets of X such that $K \subset U_y$, $y \in V_y$, and $U_y \cap V_y = \emptyset$. Now, the family $\{V_y : y \in C\}$ is an open cover of the compact set C, and hence we can find $y_1, \ldots, y_n \in C$ such that if $V = \bigcup_{i=1}^n V_{y_i}$ and $U = \bigcap_{i=1}^n U_{y_i}$, then U and V are open sets which satisfy the assertion in (d). $\square$

Corollary A.3.19

(1)   *A continuous mapping from a compact space to a Hausdorff space is a closed map (see Proposition A.2.18).*

(2)   *If $f : X \rightarrow Y$ is a continuous map, if Y is a Hausdorff space, and if $K \subset X$ is a compact subset such that $f|_K$ is 1-1, then $f|_K$ is a homeomorphism of K onto $f(K)$.*

Proof:

(1)   Closed subsets of compact sets are compact, continuous images of compact sets are compact, and compact subsets of Hausdorff spaces are closed.

(2)   This follows directly from (1) and Proposition A.2.18. $\square$

Proposition A.3.18 shows that in compact Hausdorff spaces, disjoint closed sets can be separated (just like points). There are other spaces which have this property; metric spaces, for instance, have this property, as shown by the following exercise.

Ex. A.3.20 Let (X, d) be a metric space. Given a subset $A \subset X$, define the distance from a point x to the set A by the formula

$$d(x, A) = \inf\{d(x, a) : a \in A\}. \tag{A.3.10}$$

(a) Show that $d(x, A) = 0$ if and only if x belongs to the closure $\overline{A}$ of A.
(b) Show that

$$|d(x, A) - d(y, A)| \leq d(x, y), \forall x, y \in X,$$

and hence conclude that $d : X \to \mathbb{R}$ is a continuous function.
(c) If A and B are disjoint closed sets in X, show that the function defined by

$$f(x) = \frac{d(x, A)}{d(x, A) + d(x, B)}$$

is a (meaningfully defined and uniformly) continuous function $f :$ $X \to [0, 1]$ with the property that

$$f(x) = \begin{cases} 0 & if\, x \in A \\ 1 & if\, x \in B \end{cases}. \tag{A.3.11}$$

(d) Deduce from (c) that disjoint closed sets in a metric space can be separated (by disjoint open sets). (*Hint*: if the closed sets are A and B, consider the set U (resp., V) of points where the function f of (c) is 'close to 0' (resp., 'close to 1').

The preceding exercise shows, in addition to the fact that disjoint closed sets in a metric space can be separated by disjoint open sets, that there exists lots of continuous real-valued functions on a metric space. It is a fact that the two notions are closely related. To get to this fact, we first introduce a convenient notion and then establish the relevant theorem.

Definition A.3.21 *A topological space X is said to be **normal** if (a) it is a Hausdorff space, and (b) whenever A, B are closed sets in X such that $A \cap B = \emptyset$, it is possible to find open sets U, V in X such that $A \subset U$, $B \subset V$ and $U \cap V = \emptyset$.*

The reason that we had to separately assume the Hausdorff condition in the definition given above is that the Hausdorff axiom is not a consequence of condition (b) of the preceding definition. (For example, let $X = \{1, 2\}$ and let $\tau = \{\emptyset, \{1\}, X\}$; then τ is indeed a non-Hausdorff topology, and there does not exist a pair of non-empty closed sets which are disjoint from one another so that condition (b) is vacuously satisfied.)

We first observe an easy consequence of normality, and state it as an exercise.

Ex. A.3.22   Show that a Hausdorff space X is normal if and only if it satisfies the following condition: whenever $A \subset W \subset X$, where A is closed and W is open, there exists an open set U such that $A \subset U \subset \overline{U} \subset W$. (*Hint*: consider $B = X - W$.)

Thus, we find (from Proposition A.3.18(d) and Ex. A.3.20(d)) that compact Hausdorff spaces and metric spaces are examples of normal spaces. One of the results relating normal spaces and continuous functions is the useful *Urysohn's lemma* which we now establish.

Theorem A.3.23   (Urysohn's lemma)

Suppose A and B are disjoint closed subsets of a normal space X; then there exists a continuous function $f : X \to [0, 1]$ such that

$$f(x) = \begin{cases} 0 & if \ x \in A \\ 1 & if \ x \in B. \end{cases} \tag{A.3.12}$$

Proof: Write $\mathbb{Q}_2$ for the set of 'dyadic rational numbers' in $[0,1]$. Thus, $\mathbb{Q}_2 = \{\frac{k}{2^n} : n = 0, 1, 2, \ldots, 0 \le k \le 2^n\}$.

Assertion: There exist open sets $\{U_r : r \in \mathbb{Q}_2\}$ such that

(i) $A \subset U_0, U_1 = X - B$; and
(ii)

$$r, s \in \mathbb{Q}_2, r < s \Rightarrow \overline{U_r} \subset U_s. \tag{A.3.13}$$

Proof of assertion:, Define $\mathbb{Q}_2(n) = \{\frac{k}{2^n} : 0 \le k \le 2^n\}$, for $n = 0, 1, 2, \ldots$. Then, we clearly have $\mathbb{Q}_2 = \cup_{n=0}^{\infty} \mathbb{Q}_2(n)$. We shall use induction on n to construct $U_r, r \in \mathbb{Q}_2(n)$.

First, we have $\mathbb{Q}_2(0) = \{0, 1\}$. Define $U_1 = X - B$ and appeal to Ex. A.3.22 to find an open set U_0 such that $A \subset U_0 \subset \overline{U_0} \subset U_1$. Suppose that we have constructed open sets $\{U_r : r \in \mathbb{Q}_2(n)\}$ such that the condition A.3.13 is satisfied whenever $r, s \in \mathbb{Q}_2(n)$. Notice now that if $t \in \mathbb{Q}_2(n+1) - \mathbb{Q}_2(n)$, then $t = \frac{2m+1}{2^{n+1}}$ for some unique integer $0 \le m < 2^n$. Set $r = \frac{2m}{2^{n+1}}, s = \frac{2m+2}{2^{n+1}}$, and note that $r < t < s$ and that $r, s \in \mathbb{Q}_2(n)$. Now apply Ex. A.3.22 to the inclusion $\overline{U_r} \subset U_s$ to deduce the existence of an open set U_t such that $\overline{U_r} \subset U_t \subset \overline{U_t} \subset U_s$. It is a simple matter to verify that the condition A.3.13 is satisfied now for all $r, s \in \mathbb{Q}_2(n + 1)$; the proof of the assertion is now complete.

Now define the function $f : X \to [0, 1]$ by the following prescription:

$$f(x) = \begin{cases} \inf\{t \in \mathbb{Q}_2 : x \in U_t\} & \text{if } x \in U_1 \\ 1 & \text{if } x \notin U_1 \end{cases}.$$

This function is clearly defined on all of X, takes values in $[0,1]$; it is identically equal to 0 on A, and identically equal to 1 on B. So we only need to establish the continuity of f in order to complete the proof of the theorem. We check continuity of f at a point $x \in X$. Suppose $\epsilon > 0$ is given. (In the following proof, we will use the (obvious) fact that $\mathbb{Q}_2$ is dense in $[0,1]$.)

Case (i): $f(x) = 0$: In this case, pick $r \in \mathbb{Q}_2$ such that $r < \epsilon$. The assumption that $f(x) = 0$ implies that $x \in U_r$; also $y \in U_r \Rightarrow f(y) \leq r < \epsilon$.

Case (ii): $0 < f(x) < 1$. First pick $p, t \in \mathbb{Q}_2$ such that $f(x) - \epsilon < p < f(x) < t < f(x) + \epsilon$. By the definition of f, we can find $s \in \mathbb{Q}_2 \cap (f(x), t)$ such that $x \in U_s$. Then pick $r \in \mathbb{Q}_2 \cap (p, f(x))$ and observe that $f(x) > r \Rightarrow x \notin U_r \Rightarrow x \notin \overline{U_p}$. Hence we see that $V = U_s - \overline{U_p}$ is an open neighbourhood of x; it is also easy to see that $y \in V \Rightarrow p \leq f(y) \leq s$ and hence $y \in V \Rightarrow |f(y) - f(x)| < \epsilon$.

Case (iii): $f(x) = 1$: The proof of this case is similar to part of the proof of Case (ii) and is left as an exercise for the reader.   □

We conclude this section with another result concerning the existence of 'sufficiently-many' continuous functions on a normal space.

Theorem A.3.24 (**Tietze's extension theorem**) *Suppose $f : A \to [-1, 1]$ is a continuous function defined on a closed subspace A of a normal space X. Then there exists a continuous function $F : X \to [-1, 1]$ such that $F|_A = f$.*

Proof: Let us set $f = f_0$ (for reasons that will soon become clear).

Let $A_0 = \{x \in A : f_0(x) \leq -\frac{1}{3}\}$ and $B_0 = \{x \in A : f_0(x) \geq \frac{1}{3}\}$. Then A_0 and B_0 are disjoint sets which are closed in A and hence also in X. By Urysohn's lemma, we can find a continuous function $g_0 : X \to [-\frac{1}{3}, \frac{1}{3}]$ such that $g_0(A_0) = \{-\frac{1}{3}\}$ and $g_0(B_0) = \{\frac{1}{3}\}$. Set $f_1 = f_0 - g_0|_A$ and observe that $f_1 : A \to [-\frac{2}{3}, \frac{2}{3}]$.

Next, let $A_1 = \{x \in A : f_1(x) \leq -\frac{1}{3} \cdot \frac{2}{3}\}$ and $B_1 = \{x \in A : f_1(x) \geq \frac{1}{3} \cdot \frac{2}{3}\}$; and as before, construct a continuous function $g_1 : X \to [-\frac{1}{3} \cdot \frac{2}{3}, \frac{1}{3} \cdot \frac{2}{3}]$ such that $g_1(A_1) = \{-\frac{1}{3} \cdot \frac{2}{3}\}$ and $g_1(B_1) = \{\frac{1}{3} \cdot \frac{2}{3}\}$. Then define $f_2 = f_1 - g_1|_A = f_0 - (g_0 + g_1)|_A$ and observe that $f_2 : A \to [-(\frac{2}{3})^2, (\frac{2}{3})^2]$.

Repeating this argument indefinitely, we find that we can find (i) a sequence $\{f_n\}_{n=0}^{\infty}$ of continuous functions on A such that $f_n : A \to [-(\frac{2}{3})^n, (\frac{2}{3})^n]$ for each

n, and (ii) a sequence $\{g_n\}_{n=0}^{\infty}$ of continuous functions on X such that $g_n(A) \subset [-\frac{1}{3} \cdot (\frac{2}{3})^n, \frac{1}{3} \cdot (\frac{2}{3})^n]$ for each n such that these sequences satisfy

$$f_n = f_0 - (g_0 + g_1 + \cdots + g_{n-1})|_A.$$

The series $\sum_{n=0}^{\infty} g_n$ is absolutely summable, and consequently, summable in the Banach space $C_b(X)$ of all bounded continuous functions on X. Let F be the limit of this sum. Finally, the estimate we have on f_n (see (i) above) and the equation displayed above show that $F|_A = f_0 = f$. The proof is complete. $\square$

Ex. A.3.25   (1) Show that Urysohn's lemma is valid with the unit interval $[0,1]$ replaced by any closed interval $[a, b]$ and thus justify the manner in which Urysohn's lemma was used in the proof of Tietze's extension theorem. (*Hint*: Use appropriate 'affine maps' to map any closed interval homeomorphically onto any other closed interval.) (2) Show that Tietze's extension theorem remains valid if $[0,1]$ is replaced by (a) $\mathbb{R}$, (b) $\mathbb{C}$, (c) $\mathbb{R}^n$.

A.4   The Stone–Weierstrass Theorem

We begin this section by introducing a class of spaces which, although not necessarily compact, nevertheless exhibit several features of compact spaces and are closely related to compact spaces in a manner which we shall describe. At the end of this section, we prove the very useful Stone–Weierstrass theorem concerning the algebra of continuous functions on such spaces.

The spaces we shall be concerned with are the so-called *locally compact* spaces, which we now proceed to describe.

Definition A.4.1   *A topological space X is said to be* **locally compact** *if it has a base $\mathcal{B}$ of sets with the property that given any open set U in X and a point $x \in U$, there exists a set $B \in \mathcal{B}$ and a compact set K such that $x \in B \subset K \subset U$.*

Euclidean space $\mathbb{R}^n$ is easily seen to be locally compact for every $n \in \mathbb{N}$. More examples of locally compact spaces are provided by the following proposition.

Proposition A.4.2

(1)   *If X is a Hausdorff space, the following conditions are equivalent:*

    (i)   *X is locally compact;*

    (ii)   *every point in X has an open neighbourhood whose closure is compact.*

(2)   *Every compact Hausdorff space is locally compact.*

(3)   *If X is a locally compact space, and if A is a subset of X which is either open or closed, then A, with respect to the subspace topology, is locally compact.*

Proof:

(1) $(i) \Rightarrow (ii)$: Recall that compact subsets of Hausdorff spaces are closed (see Proposition A.3.18(c)), while closed subsets of compact sets are compact in any topological space (see Proposition A.3.4(b)). It follows that if $\mathcal{B}$ is as in Definition A.4.1, then the closure of every set in $\mathcal{B}$ is compact.

$(ii) \Rightarrow (i)$: Let $\mathcal{B}$ be the class of all open sets in X whose closures are compact. For each $x \in X$, pick a neighbourhood U_x whose closure—call it K_x, is compact. Suppose now that U is any open neighbourhood of x. Let $U_1 = U \cap U_x$, which is also an open neighbourhood of x. Now, $x \in U_1 \subset U_x \subset K_x$.

Consider K_x as a compact Hausdorff space (with the subspace topology). In this compact space, we see that (a) $K_x - U_1$ is a closed, and hence compact subset of K_x, and (b) $x \notin K_x - U_1$. Hence, by Proposition A.3.18(c) we can find open sets V_1, V_2 in K_x such that $x \in V_1, K_x - U_1 \subset V_2$ and $V_1 \cap V_2 = \emptyset$. In particular, this means that $V_1 \subset K_x - V_2 = F$ (say); the set F is a closed subset of the compact set K_x and is consequently compact (in K_x and hence also in X—see Ex. A.3.3). Hence F is a closed set in (the Hausdorff space) X, and consequently, the closure in X, of V_1 is contained in F and is compact. Also, since V_1 is open in K_x, there exists an open set V in X such that $V_1 = V \cap K_x$, but since $V_1 \subset \overline{V_1} \subset F \subset U_1 \subset K_x$, we find that also $V_1 = V \cap K_x \cap U_1 = V \cap U_1$, i.e., V_1 is open in X.

Thus, we have shown that for any $x \in X$ and any open neighbourhood U of x, there exists an open set $V \in \mathcal{B}$ such that $x \in V_1 \subset \overline{V_1} \subset U$ and such that $\overline{V_1}$ is compact; thus, we have verified local compactness of X.

(2)   This is an immediate consequence of (1).

(3)   Suppose A is a closed subset of X. Let $x \in A$. Then, by (1), there exists an open set U in X and a compact subset K of X such that $x \in U \subset K$. Then $x \in A \cap U \subset A \cap K$, but $A \cap K$ is compact (in X, being a closed subset of a compact set, and hence also compact in A) and $A \cap U$ is an open set in the subspace topology of A. Thus A is locally compact.

Suppose A is an open set. Then by Definition A.4.1, if $x \in A$, then there exists sets $U, K \subset A$ such that $x \in U \subset K \subset A$, where U is open in X and K is compact. Clearly, this means that U is also open in A and we may conclude from (1) that A is indeed locally compact in this case as well.   □

In particular, we see from Proposition A.3.18(a) and Proposition A.4.2(3) that if X is a compact Hausdorff space, and if $x_0 \in X$, then the subspace $A = X - \{x_0\}$ is a locally compact Hausdorff space with respect to the subspace topology. The surprising and exceedingly useful fact is that every locally compact Hausdorff space arises in this fashion.

Theorem A.4.3   *Let X be a locally compact Hausdorff space. Then there exists a compact Hausdorff space $\hat{X}$ and a point in $\hat{X}$, which is usually referred to as the 'point at infinity' and denoted simply by ∞, such that X is homeomorphic to the subspace $\hat{X} - \{\infty\}$ of $\hat{X}$. The compact space $\hat{X}$ is customarily referred to as the* **one-point compactification** *of X.*

Proof: Define $\hat{X} = X \cup \{\infty\}$, where ∞ is an artificial point (not in X) which is adjoined to X. We make $\hat{X}$ a topological space as follows: say that a set $U \subset \hat{X}$ is open if either (i) $U \subset X$ and U is an open subset of X, or (ii) $\infty \in U$ and $\hat{X} - U$ is a compact subset of X.

Let us first verify that this prescription does indeed define a topology on $\hat{X}$. It is clear that $\emptyset$ and $\hat{X}$ are open according to our definition. Suppose U and V are open in $\hat{X}$, there are four cases:

(i) $U, V \subset X$: in this case U, V and $U \cap V$ are all open sets in X.

(ii) U is an open subset of X and $V = \hat{X} - K$ for some compact subset of X. In this case, since $X - K$ is open in X, we find that $U \cap V = U \cap (X - K)$ is an open subset of X.

(iii) V is an open subset of X and $U = \hat{X} - K$ for some compact subset of X. In this case, since $X - K$ is open in X, we find that $U \cap V = V \cap (X - K)$ is an open subset of X.

(iv)   There exists compact subsets $C, K \subset X$ such that $U = \hat{X} - C$, $V = \hat{X} - K$. In this case, $C \cup K$ is a compact subset of X and $U \cap V = \hat{X} - (C \cup K)$.

We find that in all the four cases, $U \cap V$ is open in $\hat{X}$. A similar case-by-case reasoning shows that an arbitrary union of open sets in $\hat{X}$ is also open, and so we have indeed defined a topology on $\hat{X}$.

Finally it should be obvious that the subspace topology that X inherits from $\hat{X}$ is the same as the given topology.

Since open sets in X are open in $\hat{X}$ and since X is Hausdorff, it is clear that distinct points in X can be separated in $\hat{X}$. Suppose now that $x \in X$. Then by Proposition A.4.2(1) we can find an open neighbourhood of x in X such that the closure (in X) of U is compact. If we call this closure K, then $V = \hat{X} - K$ is an open neighbourhood of ∞ such that $U \cap V = \emptyset$. Thus $\hat{X}$ is indeed a Hausdorff space.

Suppose now that $\{U_i : i \in I\}$ is an open cover of $\hat{X}$. Then, pick a U_j such that $\infty \in U_j$. Since U_j is open, the definition of the topology on $\hat{X}$ implies that

$\hat{X} - U_j = K$ is a compact subset of X. Then we can find a finite subset $I_0 \subset I$ such that $K \subset \cup_{i \in I_0} U_i$. It follows that $\{U_i : i \in I_0 \cup \{j\}\}$ is a finite sub-cover, thereby establishing the compactness of $\hat{X}$.   □

Ex. A.4.4

(1) Show that X is closed in $\hat{X}$ if and only if X is compact. (Hence if X is not compact, then X is dense in $\hat{X}$; this is the reason for calling $\hat{X}$ a compactification of X since it is 'minimal' in some sense.

(2) The following table has two columns which are labelled X and $\hat{X}$ respectively. If the ith row has spaces A and B appearing in the the first and second columns respectively, show that B is homeomorphic to $\hat{A}$.

	X	$\hat{X}$
1.	$\{1, 2, 3, \dots\}$	$\{0\} \cup \{\frac{1}{n} : n = 1, 2, \dots\}$
2.	$[0, 1)$	$[0, 1]$
3.	$\mathbb{R}^n$	$S^n = \{x \in \mathbb{R}^{n+1} : \|x\|_2 = 1\}$

We now wish to discuss continuous (real or complex-valued) functions on a locally compact space. We start with an easy consequence of Urysohn's lemma, Tietze's extension theorem and one-point compactifications of locally compact Hausdorff spaces.

Proposition A.4.5 *Suppose X is a locally compact Hausdorff space.*

(a) *If A and K are disjoint subsets of X such that A is closed and K is compact, then there exists a continuous function $f : X \to [0, 1]$ such that $f(A) = \{0\}, f(K) = \{1\}$.*

(b) *If K is a compact subspace of X, and if $f : K \to \mathbb{R}$ is any continuous function, then there exists a continuous function $F : X \to \mathbb{R}$ such that $F|_K = f$.*

Proof:

(a) In the one-point compactification $\hat{X}$, consider the subsets K and $B = A \cup \{\infty\}$. Then, $\hat{X} - B = X - A$ is open (in X, hence in $\hat{X}$). Thus B and K are disjoint closed subsets in a compact Hausdorff space, and hence, by Urysohn's lemma, we can find a continuous function $g : \hat{X} \to [0, 1]$ such that $g(B) = \{0\}, g(K) = \{1\}$. Let $f = g|_X$.

(b) This follows from applying Tietze's extension theorem (or rather, from its extension stated in Ex. A.3.25(2)) to the closed subset K of the one-point compactification of X, and then restricting the continuous function (which extends $\hat{f} : K \to \hat{X}$) to X.   □

The next result introduces a very important function space. (By the theory discussed in chapter 2, these are the most general commutative C^*-algebras.)

Proposition A.4.6   *Let $\hat{X}$ be the one-point compactification of a locally compact Hausdorff space X. Let $C(\hat{X})$ denote the space of all complex-valued continuous functions on $\hat{X}$, equipped with the sup-norm $\| \cdot \|_\infty$.*

(a)   *The following conditions on a continuous function $f : X \to \mathbb{C}$ are equivalent:*

    (i)   *f is the uniform limit of a sequence $\{f_n\}_{n=1}^\infty \subset C_c(X)$, i.e., there exists a sequence $\{f_n\}_{n=1}^\infty$ of continuous functions $f_n : X \to \mathbb{C}$ such that each f_n vanishes outside some compact subset of X with the property that the sequence $\{f_n\}$ of functions converges uniformly to f on X.*

    (ii)   *f is continuous and f vanishes at 'infinity', meaning that for each $\epsilon > 0$, there exists a compact subset $K \subset X$ such that $|f(x)| < \epsilon$ whenever $x \notin K$;*

    (iii)   *f extends to a continuous function $F : \hat{X} \to \mathbb{C}$ with the property that $F(\infty) = 0$.*

    The set of functions which satisfy these equivalent conditions is denoted by $C_0(X)$.

(b)   *Let $\mathcal{I} = \{F \in C(\hat{X}) : F(\infty) = 0\}$. Then $\mathcal{I}$ is a maximal ideal in $C(\hat{X})$ and the mapping $F \mapsto F|_X$ defines an isometric isomorphism of $\mathcal{I}$ onto $C_0(X)$.*

Proof:

(a) $(i) \Rightarrow (ii)$: If $\epsilon > 0$, pick n such that $|f_n(x) - f(x)| < \epsilon$. Let K be a compact set such that $f_n(x) = 0 \; \forall \, x \notin K$. Then, clearly, $|f_n(x)| < \epsilon \; \forall \, x \notin K$.

$(ii) \Rightarrow (iii)$: Define $F : \hat{X} \to \mathbb{C}$ by

$$F(x) = \begin{cases} f(x) & \text{if } x \in X \\ 0 & \text{if } x = \infty \end{cases};$$

the function F is clearly continuous at all points of X since X is an open set in $\hat{X}$. The continuity of F at ∞ follows from the hypothesis on f and the definition of open neighbourhoods of ∞ in $\hat{X}$ (as complements of compact sets in X).

$(iii) \Rightarrow (i)$: Fix n. Let $A_n = \{x \in \hat{X} : |F(x)| \geq \frac{1}{n}\}$ and $B_n = \{x \in \hat{X} :$ $|F(x)| \leq \frac{1}{2n}\}$. Note that A_n and B_n are disjoint closed subsets of $\hat{X}$, and that in fact $A_n \subset X$ (so that, in particular, A_n is a compact subset of X). By Urysohn's theorem we can find a continuous function $\phi_n : \hat{X} \to [0, 1]$ such that $\phi_n(A_n) = \{1\}$ and $\phi_n(B_n) = \{0\}$. Consider the function $f_n = (F\phi_n)|_X$. Then $f_n : X \to \mathbb{C}$ is continuous. Also, if we set $K_n = \{x \in \hat{X} : |F(x)| \geq \frac{1}{2n}\}$, then K_n is a compact subset of X and $X - K_n \subset B_n$, and so $f_n(x) = 0 \forall x \in X - K_n$. Finally, notice that f_n agrees with f on A_n while if $x \notin A_n$, then

$$|f(x) - f_n(x)| \leq |f(x)| (1 + |\phi_n(x)|) \leq \frac{2}{n}$$

and consequently the sequence $\{f_n\}$ converges uniformly to f.

(b)   This is obvious.   □

Now we proceed to the Stone–Weierstrass theorem via its more classical special case, the Weierstrass approximation theorem.

Theorem A.4.7   (Weierstrass approximation theorem) *The set of polynomials is dense in the Banach algebra $C[a, b]$, where $-\infty < a \leq b < \infty$.*

Proof: Without loss of generality (why?) we may restrict ourselves to the case where $a = 0, b = 1$.
    We begin with the binomial theorem

$$\sum_{k=0}^{n} \binom{n}{k} x^k y^{n-k} = (x + y)^n. \tag{A.4.14}$$

First, set $y = 1 - x$ and observe that

$$\sum_{k=0}^{n} \binom{n}{k} x^k (1 - x)^{n-k} = 1. \tag{A.4.15}$$

Next, differentiate (A.4.14) once (with respect to x, treating y as a constant) to get

$$\sum_{k=0}^{n} k \binom{n}{k} x^{k-1} y^{n-k} = n(x + y)^{n-1}; \tag{A.4.16}$$

multiply by x and set $y = 1 - x$, to obtain

$$\sum_{k=0}^{n} k \binom{n}{k} x^k (1 - x)^{n-k} = nx. \tag{A.4.17}$$

Similarly, another differentiation (of (A.4.16) with respect to x), subsequent multiplication by x^2, and the specialisation $y = 1 - x$ yields the identity

$$\sum_{k=0}^{n} k(k-1)\binom{n}{k}x^k(1-x)^{n-k} = n(n-1)x^2. \tag{A.4.18}$$

We may now deduce from the three preceding equations that

$$\sum_{k=0}^{n}(k-nx)^2\binom{n}{k}x^k(1-x)^{n-k}$$

$$= n^2x^2 - 2nx \cdot nx + (nx + n(n-1)x^2)$$

$$= nx(1-x). \tag{A.4.19}$$

In order to show that any complex-valued continuous function on $[0, 1]$ is uniformly approximable by polynomials (with complex coefficients), it clearly suffices (by considering real and imaginary parts) to show that any real-valued continuous function on $[0, 1]$ is uniformly approximable by polynomials with real coefficients.

So, suppose now that $f : [0, 1] \to \mathbb{R}$ is a continuous real-valued function and that $\epsilon > 0$ is arbitrary. Since $[0, 1]$ is compact, the function f is bounded. Let $M = \|f\|_\infty$. Also, since f is uniformly continuous, we can find $\delta > 0$ such that $|f(x) - f(y)| < \epsilon$ whenever $|x - y| < \delta$. (If you do not know what uniform continuity means, the $\epsilon - \delta$ statement given here is the definition—try to use the compactness of $[0, 1]$ and prove this assertion directly.)

We assert that if $p(x) = \sum_{k=0}^{n} f\left(\frac{k}{n}\right)\binom{n}{k}x^k(1-x)^{n-k}$, and if n is sufficiently large, then $\|f - p\| < \epsilon$. For this, first observe, thanks to (A.4.15), that if $x \in [0, 1]$ is temporarily fixed, then

$$|f(x) - p(x)| = \left|\sum_{k=0}^{n}\left(f(x) - f\left(\frac{k}{n}\right)\right)\binom{n}{k}x^k(1-x)^{n-k}\right|$$

$$\leq S_1 + S_2,$$

where $S_i = \left|\sum_{k \in I_i}\left(f(x) - f\left(\frac{k}{n}\right)\right)\binom{n}{k}x^k(1-x)^{n-k}\right|$ and the sets I_i are defined by $I_1 = \{k : 0 \leq k \leq n, |\frac{k}{n} - x| < \delta\}$ and $I_2 = \{k : 0 \leq k \leq n, |\frac{k}{n} - x| \geq \delta\}$.

Notice now that by the defining property of δ that

$$S_1 \leq \sum_{k \in I_1}\epsilon\binom{n}{k}x^k(1-x)^{n-k}$$

$$\leq \epsilon\sum_{k=0}^{n}\binom{n}{k}x^k(1-x)^{n-k}$$

$$= \epsilon,$$

while

$$S_2 \le 2M \sum_{k \in I_2} \binom{n}{k} x^k (1-x)^{n-k}$$

$$\le 2M \sum_{k \in I_2} \left(\frac{k-nx}{n\delta}\right)^2 \binom{n}{k} x^k (1-x)^{n-k}$$

$$\le \frac{2M}{n^2\delta^2} \sum_{k=0}^{n} (k-nx)^2 \binom{n}{k} x^k (1-x)^{n-k}$$

$$= \frac{2Mx(1-x)}{n\delta^2}$$

$$\le \frac{M}{2n\delta^2}$$

$$\to 0 \text{ as } n \to \infty,$$

where we used (a) the identity (A.4.19) in the 4th line above, and (b) the fact that $x(1-x) \le \frac{1}{4}$ in the 5th line above. The proof is complete.   □

We now proceed to the Stone–Weierstrass theorem, which is a very useful generalisation of the Weierstrass theorem.

Theorem A.4.8   (Stone–Weierstrass theorem)

(a) *(Real version): Let X be a compact Hausdorff space; let $\mathcal{A}$ be a (not necessarily closed) subalgebra of the real Banach algebra $C_{\mathbb{R}}(X)$; suppose that $\mathcal{A}$ satisfies the following conditions:*

   (i) *$\mathcal{A}$ contains the constant functions (or equivalently, $\mathcal{A}$ is a unital sub-algebra of $C_{\mathbb{R}}(X)$); and*
   (ii) *$\mathcal{A}$ separates points in X, meaning, of course, that if x, y are any two distinct points in X, then there exists $f \in \mathcal{A}$ such that $f(x) \ne f(y)$. Then, $\mathcal{A}$ is dense in $C_{\mathbb{R}}(X)$.*

(b) *(Complex version): Let X be as above and suppose $\mathcal{A}$ is a self-adjoint sub-algebra of $C(X)$ and suppose $\mathcal{A}$ satisfies conditions (i) and (ii) above. Then, $\mathcal{A}$ is dense in $C(X)$.*

Proof: To begin with, we may replace $\mathcal{A}$ by its closure, consequently assume that $\mathcal{A}$ is closed, and seek to prove that $\mathcal{A} = C_{\mathbb{R}}(X)$ (resp., $C(X)$).

(a)   Begin by noting that since the function $t \mapsto |t|$ can be uniformly approximated on any compact interval of $\mathbb{R}$ by polynomials (thanks to the Weierstrass' theorem), it follows that $f \in \mathcal{A} \Rightarrow |f| \in \mathcal{A}$. Since $x \vee y = \max\{x, y\} = \frac{x+y+|x-y|}{2}$ and $x \wedge y = \min\{x, y\} = \frac{x+y-x\vee y}{2}$, it follows that $\mathcal{A}$ is a 'sub-lattice' of $C_\mathbb{R}(X)$ meaning that $f, g \in \mathcal{A} \Rightarrow f \vee g$, $f \wedge g \in \mathcal{A}$.

Next, the hypothesis that $\mathcal{A}$ separates points of X (together with the fact that $\mathcal{A}$ is a vector space containing the constant functions) implies that if x, y are distinct points in X and if $s, t \in \mathbb{R}$ are arbitrary, then there exists $f \in \mathcal{A}$ such that $f(x) = s$, $f(y) = t$. (Reason: first find $f_0 \in \mathcal{A}$ such that $f_0(x) = s_0 \neq t_0 = f_0(y)$; next, find constants $a, b \in \mathbb{R}$ such that $as_0 + b = s$ and $at_0 + b = t$, and set $f = af_0 + b1$, where, of course, 1 denotes the constant function 1.)

  Suppose now that $f \in C_\mathbb{R}(X)$ and that $\epsilon > 0$. Temporarily, fix $x \in X$. For each $y \in X$, we can, by the previous paragraph, pick $f_y \in \mathcal{A}$ such that $f_y(x) = f(x)$ and $f_y(y) = f(y)$. Next, choose an open neighbourhood U_y of y such that $f_y(z) > f(z) - \epsilon \; \forall \, z \in U_y$. Then, by compactness, we can find $\{y_1, \dots, y_n\} \subset X$ such that $X = \cup_{i=1}^n U_{y_i}$. Set $g^x = f_{y_1} \vee f_{y_2} \vee \cdots \vee f_{y_n}$, and observe that $g^x(x) = f(x)$ and that $g^x(z) > f(z) - \epsilon \; \forall \, z \in X$.

Now, we can carry out the procedure outlined in the preceding paragraph, for each $x \in X$. If g^x is as in the last paragraph, then, for each $x \in X$, find an open neighbourhood V_x of x such that $g^x(z) < f(z) + \epsilon \; \forall \, z \in V_x$. Then, by compactness, we may find $\{x_1, \dots, x_m\} \subset X$ such that $X = \cup_{j=1}^m V_{x_j}$. Finally, set $g = g^{x_1} \wedge g^{x_2} \wedge \cdots \wedge g^{x_m}$, and observe that the construction implies that $f(z) - \epsilon < g(z) < f(z) + \epsilon \; \forall \, z \in X$, thereby completing the proof of (a).

(b) This follows easily from (a), upon considering real and imaginary parts. (This is where we require that $\mathcal{A}$ be a self-adjoint sub-algebra in the complex case.) □

We state some useful special cases of the Stone–Weierstrass theorem in the exercise below.

Ex. A.4.9 In each of the following cases, show that the algebra $\mathcal{A}$ is dense in $C(X)$:

(i) $X = \mathbb{T} = \{z \in \mathbb{C} : |z| = 1\}$, $\mathcal{A} = \bigvee\{z^n : n \in \mathbb{Z}\}$ and thus, $\mathcal{A}$ is the class of 'trigonometric polynomials'.

(ii) X a compact subset of $\mathbb{R}^n$ and $\mathcal{A}$ is the set of functions of the form $f(x_1, \dots, x_n) = \sum_{k_1, \dots, k_n = 0}^N \alpha_{k_1, \dots, k_n} x_1^{k_1} \cdots x_n^{k_n}$, where $\alpha_{k_1, \dots, k_n} \in \mathbb{C}$.

(iii) X a compact subset of $\mathbb{C}^n$, and $\mathcal{A}$ is the set of (polynomial) functions of the form

$$f(z_1, \ldots, z_n) = \sum_{k_1.l_1,\ldots,k_n.l_n=0}^{N} \alpha_{k_1.l_1,\ldots,k_n.l_n} z_1^{k_1} \overline{z_1}^{l_1} \cdots z_n^{k_n} \overline{z_n}^{l_n}.$$

(iv) $X = \{1, 2, \ldots, N\}^{\mathbb{N}}$, and $\mathcal{A} = \{\omega_{k_1,\ldots,k_n} : n \in \mathbb{N}, 1 \leq k_1, \ldots, k_n \leq N\}$, where

$$\omega_{k_1,\ldots,k_n}((x_1, x_2, \ldots)) = \exp\left(\frac{2\pi i \sum_{j=1}^{n} k_j x_j}{N}\right).$$

The 'locally compact' version of the Stone–Weierstrass theorem is the content of the next exercise.

Ex. A.4.10 Let $\mathcal{A}$ be a self-adjoint sub-algebra of $C_0(X)$, where X is a locally compact Hausdorff space. Suppose $\mathcal{A}$ satisfies the following conditions:

(i)   if $x \in X$, then there exists $f \in \mathcal{A}$ such that $f(x) \neq 0$; and
(ii) $\mathcal{A}$ separates points.

Then show that $\mathcal{A} = C_0(X)$. (*Hint*: Let $\mathcal{B} = \{F + \alpha 1 : f \in \mathcal{A}, \alpha \in \mathbb{C}\}$, where 1 denotes the constant function on the one-point compactification $\hat{X}$, and F denotes the unique continuous extension of f to $\hat{X}$. Appeal to the already established compact case of the Stone–Weierstrass theorem.)

B

Tables

Table B. 1   Characteristic functions of standard distributions

Distribution	Characteristic Function $\phi(t), t \in \mathbb{R}$
Bernoulli (p)	$1 - p + pe^{it}$
Binomial (n, p)	$(1 - p + pe^{it})^n$
Uniform $(\{1, 2, \ldots, n\})$	$\dfrac{e^{it}(1-e^{it})}{n(1-e^{int})}$
Poisson (λ)	$e^{\lambda}(e^{iu-1})$
Uniform (a, b)	$\dfrac{e^{ibt} - e^{iat}}{i(b-a)t}$
Normal (m, σ^2)	$e^{-imt - t^2 \frac{\sigma^2}{2}}$

Table B. 2   Normal tables evaluating $\frac{1}{2\pi} \int_0^z e^{-\frac{x^2}{2}} dx$

z	0.00	0.01	0.02	0.03	0.04	0.05	0.06	0.07	0.08	0.09
0.0	0.0000	0.0040	0.0080	0.0120	0.0160	0.0199	0.0239	0.0279	0.0319	0.0359
0.1	0.0398	0.0438	0.0478	0.0517	0.0557	0.0596	0.0636	0.0675	0.0714	0.0753
0.2	0.0793	0.0832	0.0871	0.0910	0.0948	0.0987	0.1026	0.1064	0.1103	0.1141
0.3	0.1179	0.1217	0.1255	0.1293	0.1331	0.1368	0.1406	0.1443	0.1480	0.1517
0.4	0.1554	0.1591	0.1628	0.1664	0.1700	0.1736	0.1772	0.1808	0.1844	0.1879
0.5	0.1915	0.1950	0.1985	0.2019	0.2054	0.2088	0.2123	0.2157	0.2190	0.2224
0.6	0.2258	0.2291	0.2324	0.2357	0.2389	0.2422	0.2454	0.2486	0.2518	0.2549
0.7	0.2580	0.2612	0.2642	0.2673	0.2704	0.2734	0.2764	0.2794	0.2823	0.2852
0.8	0.2882	0.2910	0.2939	0.2967	0.2996	0.3023	0.3051	0.3079	0.3106	0.3133
0.9	0.3159	0.3186	0.3212	0.3238	0.3264	0.3290	0.3315	0.3340	0.3365	0.3389
1.0	0.3414	0.3438	0.3461	0.3485	0.3508	0.3531	0.3554	0.3577	0.3599	0.3622
1.1	0.3643	0.3665	0.3687	0.3708	0.3729	0.3749	0.3770	0.3790	0.3810	0.3830
1.2	0.3849	0.3869	0.3888	0.3907	0.3925	0.3944	0.3962	0.3980	0.3997	0.4015
1.3	0.4032	0.4049	0.4066	0.4083	0.4099	0.4115	0.4131	0.4147	0.4162	0.4177
1.4	0.4193	0.4207	0.4222	0.4237	0.4251	0.4265	0.4279	0.4292	0.4306	0.4319
1.5	0.4332	0.4345	0.4358	0.4370	0.4382	0.4394	0.4406	0.4418	0.4430	0.4441
1.6	0.4452	0.4463	0.4474	0.4485	0.4495	0.4505	0.4516	0.4526	0.4535	0.4545
1.7	0.4554	0.4564	0.4573	0.4582	0.4591	0.4600	0.4608	0.4617	0.4625	0.4633
1.8	0.4641	0.4649	0.4656	0.4664	0.4671	0.4679	0.4686	0.4693	0.4700	0.4706
1.9	0.4713	0.4720	0.4726	0.4732	0.4738	0.4744	0.4750	0.4756	0.4762	0.4767
2.0	0.4773	0.4778	0.4783	0.4788	0.4793	0.4798	0.4803	0.4808	0.4813	0.4817
2.1	0.4822	0.4826	0.4830	0.4834	0.4838	0.4842	0.4846	0.4850	0.4854	0.4858
2.2	0.4861	0.4865	0.4868	0.4871	0.4875	0.4878	0.4881	0.4884	0.4887	0.4890
2.3	0.4893	0.4896	0.4898	0.4901	0.4904	0.4906	0.4909	0.4911	0.4914	0.4916
2.4	0.4918	0.4920	0.4923	0.4925	0.4927	0.4929	0.4931	0.4933	0.4934	0.4936
2.5	0.4938	0.4940	0.4942	0.4943	0.4945	0.4946	0.4948	0.4949	0.4951	0.4952
2.6	0.4954	0.4955	0.4956	0.4957	0.4959	0.4960	0.4961	0.4962	0.4963	0.4964
2.7	0.4966	0.4967	0.4968	0.4969	0.4969	0.4970	0.4971	0.4972	0.4973	0.4974
2.8	0.4975	0.4975	0.4976	0.4977	0.4978	0.4978	0.4979	0.4980	0.4980	0.4981
2.9	0.4982	0.4982	0.4983	0.4983	0.4984	0.4984	0.4985	0.4985	0.4986	0.4986
3.0	0.4987	0.4987	0.4988	0.4988	0.4988	0.4989	0.4989	0.4989	0.4990	0.4990
3.1	0.4991	0.4991	0.4991	0.4991	0.4992	0.4992	0.4992	0.4993	0.4993	0.4993
3.2	0.4993	0.4994	0.4994	0.4994	0.4994	0.4994	0.4995	0.4995	0.4995	0.4995
3.3	0.4995	0.4996	0.4996	0.4996	0.4996	0.4996	0.4996	0.4996	0.4997	0.4997
3.4	0.4997	0.4997	0.4997	0.4997	0.4997	0.4997	0.4997	0.4998	0.4998	0.4998

References

[1] W. Arveson, *An Invitation to C*-Algebras*, Springer, New York

[2] R. B. Ash, *Probability and Measure Theory*, 2d ed., with contributions from Catherine Dolans-Dade, Harcourt/Academic Press, Burlington, MA, 2000.

[3] P. Billingsley, *Probability and Measure*, New York, John Wiley, 1986.

[4] P. Billingsley, *Convergence of Probability Measures*, 2d ed., Wiley Series in Probability and Statistics: Probability and Statistics, John Wiley & Sons, Inc., New York, 1999.

[5] L. Brieman, *Probability*, Society for Industrial and Applied Mathematics, Philadelphia, 1993.

[6] V. S. Borkar, *Probability Theory, An Advanced Course*, Universitext, Springer-Verlag, New York, 1995.

[7] K. L. Chung, *A Course in Probability Theory*, 3d ed., Academic Press, Inc., San Diego, CA, 2001.

[8] D. Dawson, *Introduction to Markov Chains*, from lectures delivered to the Twelfth Biennial Seminar of the Canadian Mathematical Congress (Vancouver, B.C., 1969), Canadian Mathematical Monographs, No. 2 Canadian Mathematical Congress, Montreal, Que., 1969

[9] J. L. Doob, *Measure Theory: Graduate Texts in Mathematics*, 143, Springer-Verlag, New York, 1994.

[10] R. Durrett, *Essentials of Stochastic Processes*, Springer Texts in Statistics, Springer-Verlag, New York, 1999.

[11] W. Feller, *An Introduction to Probability Theory and Its Applications*, Vol. I, 3d ed., John Wiley & Sons, Inc., New York-London-Sydney, 1968.

[12] W. Feller, *An Introduction to Probability Theory and Its Applications*, Vol. II, 2d ed., John Wiley & Sons, Inc., New York-London-Sydney, 1971.

[13] M. I. Friedlin and A. D. Wentzell, *Random Perturbations of Dynamical Systems*, Springer-verlag, New York, 1998.

[14] Paul R. Halmos, *Measure Theory*, Van Nostrand, Princeton, 1950.

[15]   P. G. Hoel, S. C. Port, and C. J. Stone, *Introduction to Stochastic Processes*, the Houghton Mifflin Series in Statistics, Houghton Mifflin Co., Boston, Mass., 1972

[16]   J. Jacod and P. Protter *Probability Essentials*, 2d ed., Universitext, Springer-Verlag, Berlin, 2003.

[17]   S. Karlin and H. M. Taylor, *A First Course in Stochastic Processes*, 2d ed., Academic Press, New York-London, 1975.

[18]   M. Keane, *The Essence of Large Numbers*, Algorithms, Fractals, and Dynamics (Okayama/Kyoto, 1992), 125–129, Plenum, New York, 1995.

[19]   J. L. Kelley, *General Topology*, Reprint of the 1955 edition [Van Nostrand, Toronto, Ont.], *Graduate Texts in Mathematics*, No. 27. Springer-Verlag, New York-Berlin, 1975.

[20]   J. F. C. Kingman, and S. J. Taylor, *Introduction to Measure and Probability*, Cambridge University Press, London-New York-Ibadan, 1966.

[21]   M. Loéve, *Probability Theory I*, 4 ed., *Graduate Texts in Mathematics*, Vol. 45. Springer-Verlag, New York-Heidelberg, 1977.

[22]   M. Loéve, *Probability Theory II*, 4 ed., *Graduate Texts in Mathematics*, Vol. 46. Springer-Verlag, New York-Heidelberg, 1977.

[23]   F. A. Murray and John von Neumann, On rings of operators, *Ann. Math.*, 37, (1936), 116-229.

[24]   K. R. Parthasarathy, *Introduction to Probability and Measure*, Texts and Readings in Mathematics, 33. Hindustan Book Agency, New Delhi, 2005.

[25]   H. L. Royden, *Real Analysis*, 3d ed., Macmillan Publishing Company, New York, 1988.

[26]   Walter Rudin, *Real and Complex Analysis*, 3d ed., McGraw-Hill Book Company, New Delhi 1987.

[27]   George F. Simmons, *Topology and Modern Analysis*, McGraw-Hill Book Company, 1963.

[28]   V. S. Sunder, *Functional Analysis : Spectral Theory*, Birkhaüser, Basel, 1997.

[29]   V. S. Varadarajan, *On a theorem of F. Riesz concerning the form of linear functionals*, Fund. Math., **46**, (1959), 209–220.

Index

2^{Ω}, 3
E', 3
$L^p(\Omega, \mathcal{B}, \mu)$, 142
$\mathcal{L}^0$, 33
$\mathcal{L}^0_+$, 33
$\mathcal{S}_+$, 29
$\mathcal{L}^1(\Omega, \mathcal{B}, \mu)$, 36
$\mathcal{M}(\mu)$, 13
ϵ-net, 194
$\liminf f_n$, 31
$\limsup f_n$, 31
μ^*-measurable, 13
$\sigma(\mathcal{S})$, 3
σ-algebra, 3
σ-finite measure, 16
n-dimensional cylinder set, 24
x-slice, 57
y-slice, 57
$\mathcal{M}(\mathcal{S})$, 7
$\mathcal{A}(\mathcal{S})$, 7

a.e., 39
absolutely continuous, 149
absolutely continuous function, 43,165
Alexander sub-base theorem, 198
algebra, 6

Bernoulli trials, 23
Borel σ- algebra, 5
Borel–Cantelli lemma, 45
Borel sets, 5
bounded linear functional, 147
bounded variation, 163
Brownian motion, 72

Cantor set, 20
Caratheodory extension theorem, 16
Cauchy–Schwarz inequality, 147
central limit theorem, 90
chain rule for derivatives, 151
Chapman–Kolmogorov equations, 71
closed, 186
closed map, 192
closure, 187
compact, 193
completion of measure space, 18
complex measures, 137
conditional probability, 53, 71
conjugate indices, 143
continuity theorem, 88
continuous, 189
continuous from above, 10
continuous from below, 10
countably additive, 9
countably sub-additive, 10
counting measure, 11

dense, 187
diagonal argument, 196
differentiable, 160
directed set, 185
distribution, 42
    Beta $(k, n - k + 1)$, 52
    Binomial (n, p), 48
    Brenoulli (p), 47
    Gamma (n, λ), 52
    Geometric (p), 48
    Hypergeometric (N, n, g), 49
    Negative Binomial (r, p), 49

Normal (m, σ^2), 51
Poisson (λ), 49
Uniform (a, b), 50
Uniform $\{1, 2, 3, \ldots, n\}$, 48
distribution function, 21
dominated convergence theorem, 37
dual space, 148

Egoroff's theorem, 82
Ergodic theorem, 100

Fatou's lemma, 35
finite complex measure, 137
finite intersection property, 193
finite measure, 9
finite measure space, 17
finite real measure, 137
finitely additive, 8
Fubini's theorem, 58

Hölder's inequality, 143
Hahn decomposition, 152
Hilbert space, 147
homeomorphic, 192
homeomorphism, 192

i.i.d, 87
indefinite integral, 41, 148
independence, 45, 61
inner regularity, 23
integrable, 36
invariant σ-algebra, 92
inversion theorem, 75

joint distribution, 61

Kolmogorov consistency theorem, 69
Kolmogorov's zero–one law, 47

Lebesgue measurable set, 20
Lebesgue measure, 4
Lebesgue–Nikodym theorem, 153

Markov chain, 101
    kth return time, 124
    accessible, 108

aperiodic, 108
closed communicating class, 108
communicate, 108
Ergodic theorem, 132
hitting time, 109
inter-arrival times, 124
irreducible, 108
Lyapunov function, 115
null recurrent, 133
period, 108
positive recurrent, 133
recurrent, 109
Stationary distribution, 128
stopping time, 123
transient, 109
transition matrix, 101
Markov processes, 70
MCT, 35
measurable, 26
measurable rectangle, 56
measurable space, 17
measure, 9
measure space, 17
metric, 182
metric space, 182
monotone class, 6
monotone class lemma, 6
monotone convergence theorem, 35
mutually absolutely continuous, 151
mutually singular, 152

net, 185
norm, 182
normed vector space, 183

one-point compactification, 207
open, 184
open cover, 193
open map, 192
outer measure, 12
outer regularity, 23

Polya's theorem, 84
probability measure, 9

probability measure space, 17
product measure, 56
product measure space, 56

Radon–Nikodym derivative, 151
Radon–Nikodym theorem, 148
random variable
    absolutely continuous, 43
    almost everywhere convergence, 79
    characteristic function, 73
    conditional expectation, 53
    continuous, 42
    convergence in rth mean, 79
    convergence in distribution, 79
    convergence in probability, 79
    discrete, 42
    expectation, 43
    moment generating function, 44
    weak convergence, 79
random walk, 120
reflection principle, 127
regularity, 23
Riesz lemma, 148
Riesz representation theorem, 168

Scheffe's theorem, 86
second axiom of countability, 196
separable, 196
set function, 8
simple function, 27
Skorokhod's theorem, 83
stationary sequence, 92
stochastic matrix, 66
Stone–Weierstrass theorem, 212

strong law of large numbers, 94
strong Markov property, 123

tail σ-algebra, 47
tail event, 47
Tchebychev's inequality, 43
Tietze's extension theorem, 204
Tonelli's theorem, 57
topological space, 184
    Hausdorff, 199
    locally compact, 205
    normal, 202
topology, 184
    base, 188
    co-countable, 185
    co-finite, 184
    discrete, 184
    indiscrete, 184
    subspace, 191
    weak, 190
    product, 191
    sub-base, 188
total variation measure, 138
total variation of function, 163
totally bounded, 194
translation invariant measure, 18
Tychonoff's theorem, 199

uniqueness theorem, 75
Urysohn's lemma, 203

Vitali cover, 159

weak law of large numbers, 93
Weierstrass approximation theorem, 210

Printed and bound by CPI Group (UK) Ltd, Croydon, CR0 4YY

18/10/2024

01776259-0020